Aspects of Microbiology 9

Extracellular Enzymes

Fergus G. Priest

 Van Nostrand Reinhold (UK) Co. Ltd

Fergus Priest is currently a Lecturer in Microbiology at Heriot-Watt University, Edinburgh. He is a member of the Society for General Microbiology, Society for Applied Bacteriology, Institute of Brewing and the American Society for Microbiology. His research interests include the regulation of extracellular enzyme synthesis, protein secretion in Gram-positive bacteria, and novel enzymes of biotechnological potential.

This edition is not for sale in the USA

First published in 1984 by
Van Nostrand Reinhold (UK) Co. Ltd
Molly Millars Lane, Wokingham, Berkshire, England

Photoset in Times 9 on 10pt by Kelly Typesetting Limited
Bradford-on-Avon, Wiltshire

Printed and bound in Hong Kong

ISBN 0 442 30588 5

Preface

The experience gained over the past 50 or 60 years in the manufacture of industrial enzymes from microorganisms encompasses microbiology, biochemistry and chemical engineering and is assuming considerable importance in the development of biotechnology. Most industrially important enzymes are extracellular. A detailed picture of the molecular biology of protein secretion is emerging from the concerted efforts of both biochemists and geneticists and it is becoming apparent that the secretion process is essentially the same in both prokaryotic and eukaryotic cells. These studies are providing the exciting prospect of the secretion of foreign proteins, such as insulin, by microbial cells containing cloned genes. The commercial implications include simple and efficient product recovery and increased yields. Moreover, yields of extracellular enzymes have been steadily improved by genetic manipulation. Although this has not yet provided a detailed understanding of the regulation of extracellular enzyme synthesis—largely due to the lack of exploitable genetic systems for the relevant microorganisms—the approaches that have been adopted should be of interest and value to those involved in the manufacture of a variety of microbial products.

Scale-up of laboratory scale procedures to the pilot plant and into commercial practice is causing considerable problems in many microbial processes. Here again, the wealth of experience gained from producing bacterial and fungal enzymes on an industrial scale should be invaluable to those venturing into similar microbiologically based industries.

Finally, the enzyme industry itself is worthy of attention. From its inception early this century it has expanded into food technology, waste product utilization and pharmaceuticals. Two major boosts to the industry were the inclusion of alkaline proteases from *Bacillus* strains in household washing detergents in the 1960s, and the development of enzymes immobilized on solid supports later that decade. Immobilized glucose isomerase is used for the conversion of glucose (derived from the enzymic hydrolysis of starch) into the sweeter-tasting fructose. The replacement of sucrose in many foods and beverages by these high fructose corn syrups has helped to promote the enzyme industry into multi-million dollar markets which promise to expand further as new enzymes are discovered and processes invented.

This book deals with both the commercial and academic aspects of extracellular enzyme synthesis. It describes those enzymes that are produced on an industrial scale and outlines their uses and how they are manufactured. It also provides detailed coverage of the molecular biology of protein secretion, the regulation of protein synthesis and current approaches to increasing enzyme yield. It should therefore be of value to advanced undergraduate and graduate students in microbiology, biochemistry and related disciplines who are seeking a concise account of this branch of industrial microbiology. It should also provide an up-to-date and straightforward account of the molecular biology of extracellular enzyme synthesis for those involved in the microbiologically based industries. Thus it is an attempt to bridge one of those gaps between academic and industrial microbiology that now comes under the umbrella of biotechnology.

1984   F. G. PRIEST, *Edinburgh*

Contents

Contents

1 Introduction

Microorganisms are responsible for the recycling of much of the organic material in the environment. As animals and plants die, they are attacked by small animals and microorganisms. Their constituent molecules are liberated and used by these saprophytes as a supply of energy and to make new cell components. The low-molecular weight, water-soluble materials are readily assimilated but most of the original organism comprises macromolecules. In plants, cellulose, hemicellulose, lignin, pectin and starch predominate; in animals, proteins, glycoproteins, glycogen and chitin are major constituents. Microorganisms contain specialized cell wall polymers such as peptidoglycan and all organisms contain nucleic acids. These macromolecules are a major food source for heterotrophic microorganisms but, since they are so large, they cannot be readily utilized. Microorganisms have adopted essentially two strategies to enable them to metabolize these compounds. The compound can be engulfed by the cytoplasmic membrane to form a vacuole within the cytoplasm. Enzymes are secreted into this vacuole and the polymeric substrates degraded and subsequently metabolized. Uptake of water and aqueous solutions by this method is referred to as pinocytosis; uptake of particulate matter is termed phagocytosis. Since the prokaryotic membrane is unable to carry out these processes, pinocytosis and phagocytosis are restricted to those eukaryotic microorganisms that lack a cell wall. The major group of such organisms is the protozoa. Those eukaryotes and prokaryotes that possess a cell wall have adopted an alternative strategy for the assimilation of macromolecular nutrients. Enzymes are liberated by the cell, or colony of cells, degrade polymeric material in the environment and the low-molecular weight products are assimilated. Consequently, extracellular enzymes are common in those microorganisms that inhabit soil and decaying plant and animal matter. Amongst the bacteria, strains of *Bacillus, Clostridium, Cytophaga* and many actinomycetes, in particular streptomycetes are prolific producers of extracellular enzymes. Moreover, Gram negative bacteria such as vibrios, aeromonads and pseudomonads are common in decomposing seaweeds and other marine habitats and often secrete agarases and similar enzymes. Filamentous fungi and yeasts also secrete a variety of extracellular enzymes.

It will be apparent that most extracellular enzymes are depolymerases acting on polysaccharides, proteins and nucleic acids. The majority are hydrolases, although exceptions do occur such as the pectin lyases which are in fact *trans* eliminases (see Chapter 4). Some extracellular enzymes have low-molecular weight substrates. A notable example is penicillinase (β-lactamase) which hydro-lyses the β-lactam ring of penicillin and renders the antibiotic harmless. Since penicillin inhibits cell wall synthesis, it is essential that it should be inactivated in the environment before it can bind to the cell surface.

It will be useful at this point to define the term 'extracellular' as it pertains to enzymes, since this term has caused confusion in the past. It is now generally agreed that extracellular refers to any enzyme that crosses the cytoplasmic membrane. Stricly speaking, digestive enzymes within the phagocytic vacuole of the protozoa, or enzymes liberated into the environment by a bacterium, are all extracellular since they have crossed the cytoplasmic membrane. The final location of an extracellular enzyme will therefore be determined by the structure of the cell.

Cell wall structure and enzyme location

Bacteria are traditionally divided into two groups depending on their reaction to the Gram strain. This in turn reflects the chemical composition and structure of the cell wall (Rogers, 1983). Gram positive bacteria have a relatively simple cell wall comprising a thick coat (about 20 nm) of peptidoglycan containing covalently bound teichoic acid. This net-like molecule bounds the cytoplasmic membrane and provides structural strength to the cell. Extracellular enzymes cross the membrane and may be temporarily restricted by the cell wall but eventually diffuse into the environment. Some enzymes, however, remain attached to the outer surface of the membrane. Since these molecules have crossed the membrane they are considered to be extracellular but, on forming protoplasts (by enzymic removal of the cell wall in isotonic medium) they are partially or completely released from the membrane. Under certain growth conditions such enzymes may be naturally released from the cell, the alkaline phosphatase and α-glucosidase of *Bacillus licheniformis* being two examples. A third location for an enzyme in the Gram positive cell is anchored to the inner surface of the membrane. Strictly speaking this is not an extracellular location since the enzyme does not traverse the membrane.

The envelope of Gram negative bacteria is a complicated structure comprising two membranes (Rogers, 1983). The cytoplasmic membrane is bounded by a thin layer of peptidoglycan. This is surrounded by the outer membrane and between these two hydrophobic barriers lies a hydrophilic space, the periplasm. The periplasm may account for 20 to 40% of the total cell mass and contains a variety of proteins including specific amino acid and sugar binding proteins and hydrolytic enzymes. There are therefore several locations for enzymes in Gram negative bacteria: cytoplasmic, anchored to the inside or outside of the cytoplasmic membrane, in the periplasm, fixed to the inner or outer surface of the outer membrane, or secreted into the environment. Since all proteins except those in the cytoplasm or on the inner surface of the cytoplasmic membrane have crossed the membrane and are released by osmotic shock treatment or by conversion of the cells to sphaeroplasts (osmotically fragile cells derived by lysozyme treatment), these enzymes are considered to be extracellular. Many of the conventional extracellular enzymes of Gram positive bacteria may have their counterparts in the periplasmic enzymes of Gram negative bacteria. Consequently extracellular enzymes *sensu stricto* are relatively rare in Gram negative bacteria but do occur particularly in pseudomonads, aeromonads and some enterobacteria. Indeed, the enterotoxin of *Vibrio cholerae* is secreted into the surrounding medium.

Structurally the fungal cell wall resembles the Gram positive bacterial wall. It is largely comprised of 1,3-α- and 1,3-β-glucan with chitin and varying amounts of cellulose and protein. It is not, however, a homogeneous mixture of these constituent polymers but appears to be a structured and complex assembly. Although little is known about the secretion of proteins by fungi, it is generally assumed that the molecules diffuse through the wall once released from the cytoplasm. With regard to enzyme secretion, the fungal cell wall is therefore similar to the Gram positive bacterial wall.

Commercial enzymes

The exploitation of enzymes is not a recent development: they have been used throughout the ages in leather tanning and cheese making, in the preparation of malted barley for beer brewing and in the leavening of bread. These processes used enzymes in the form of animal and plant tissues or whole microorganisms. The birth of commercial enzymes as partially-purified preparations from living cells is more recent, and can be traced to the end of the last century. Jokichi Takamine, a Japanese scientist living in the USA, filed the first patent for an enzyme in 1894. In this process, *Aspergillus oryzae* was grown on moist rice or wheat bran and the secreted amylase was extracted with water or salts. This 'Takadiastase' is still used as a digestive aid today. The use of bacteria, in particular *Bacillus* strains, for enzyme production followed some twenty years later and again involved growing the microorganism as a surface pellicle in trays of semi-solid medium. The usefulness of extracellular enzymes was readily appreciated. Extracellular enzymes are easier to recover and purify than their cytoplasmic counterparts; in particulr, cell breakage is unnecessary and problems involving removal of nucleic acids are absent. Secondly, it is easier to obtain very high yields of extracellular enzymes because the yield is not restricted by the biomass obtainable. Consequently, the submerged culture techniques developed by the antibiotic industry in the 1940s were readily adopted by enzyme manufacturers, and the increase in productivity provided by improved control of growth conditions boosted the industry considerably. There followed slow but steady growth in the 1950s that was dramatically amplified the following decade by the introduction of enzyme washing powders containing alkaline protease from *Bacillus licheniformis*. Microbial rennets for cheese manufacture and the enzymic conversion of starch into a mixture of glucose and fructose for use as a food sweetener have since been developed and represent two recent growth areas of industrial enzyme usage. Furthermore, current interest in the efficient utilization of renewable resources and the pressure on industry to work within environmentally acceptable limits has stimulated wider interest in enzymes. These factors have combined to produce a world market for industrial enzymes in 1981 variously estimated at between $150 million and $400 million. It is predicted that this will rise to some $600 million by 1985. The bulk of this market comprises proteases and carbohydrases which together account for about 90% of the total; the remainder includes technical and pharmaceutical products. Considering the thousands of enzymes known, this is a very restricted sample but it emphasises that it is much easier to discover a new enzyme than to identify a profitable market.

This book deals with the industrial enzymes (Table 1). The production trend over the past 30 years has been away from animal and plant sources and towards microorganisms to the extent that new products are almost invariably derived from bacteria or fungi. There are several reasons for this: (1) microorganisms grow rapidly and are ideal for intensive cultivation, (2) medium constituents are cheap and generally comprise agricultural products available in bulk and (3) choice of producer-organism is wide and can be improved by genetic manipulation. Thus the often variable and unpredictable sources of animal and plant enzymes are

Table 1 Commercial enzyme products

Source/name	Commercially available before: 1900	1950	1980	Production in 1980: tonnes/year	as % of total
Animal					
Rennet	X			2	0.15
Trypsin		X		15	1
Pepsin		X		5	0.4
Plant					
Papain		X		100	8
Microbial					
Fungal amylase	X			10	0.8
Bacterial amylase		X		300	23
Glucoamylase			X	300	23
Fungal protease	X			10	0.8
Bacillus protease		X		500	38
Pectinase		X		10	0.8
Glucose isomerase			X	50	4
Fungal rennet			X	10	0.8

gradually being replaced by microbial equivalents, although for some applications animal and plant enzymes have retained their market share.

In this book, those enzymes produced on a commercial scale from microorganisms and their uses are described. However, the field of extracellular enzymes is approaching a revolutionary phase. As more is learnt of the process of protein secretion across membranes and the techniques of genetic engineering become more sophisticated, the prospect emerges that virtually any protein may be made extracellular. Thus, a process that originated as a means of scavenging nutrients from the environment will be exploited on a large scale to engineer microorganisms that can secrete high yields of valuable proteins. To understand how this will come about, we must first consider the process of protein secretion and the regulation of protein synthesis. This will be followed by an account of the current enzyme products and their uses. The penultimate chapters will focus on screening strains for potentially useful enzymes and the development of genetically engineered strains that secrete large amounts of protein. Finally, some engineering aspects of industrial scale production and purification of proteins will be considered.

Reference

ROGERS, H. J. (1983). *Bacterial Cell Structures*. Van Nostrand Reinhold (UK), England.

2 Enzyme secretion

Signal hypothesis

The key feature of an extracellular enzyme is that it is transported across a membrane. The central question to the understanding of protein export is therefore, how does the cell distinguish between cytoplasmic proteins and those destined either for incorporation into the membrane or across it to some other location? Precursor forms of secretory proteins usually contain an NH_2-terminal extension of some 15 to 32 amino acids. It has been proposed that this 'signal' or 'leader' peptide, as it emerges from the ribosome, directs the ribosome to the membrane. According to the original model, the signal peptide recruited other membrane proteins to form a pore or tunnel in the membrane that was stabilized by attachment of the ribosome. As the protein was synthesized, it was exported through this pore in a process termed cotranslational secretion. Once part, or all, of the protein had been exported, the signal peptide was removed by a specific protease (signal peptidase). The elements of the signal hypothesis are shown in Figure 1, and, although it has been substantially refined in recent years the underlying theme remains correct.

Five important principles have emerged from the signal hypothesis: (1) there is no difference beteen membrane bound and cytoplasmic ribosomes, (2) secreted proteins are generally synthesized in a larger, precursor form, (3) proteins are secreted cotranslationally (although not invariably), (4) there can be a stop-secretion sequence within the protein that halts secretion giving rise to an integral, membrane protein, (5) some form of export machinery in the membrane is required and (6) the process of protein export has been highly conserved throughout evolution to the extent that prokaryotic cells recognize eukaryotic signal sequences and *vice versa*. These aspects of protein secretion will be examined in detail in this chapter.

Precursor forms of exported proteins   Following formulation of the signal hypothesis, nascent secretory proteins from various systems were characterized and the signal peptides analysed. This was fairly straight-forward since the precursor form of the protein was generally larger than the mature form and could be separated from the mature protein by sodium dodecyl sulphate-polyacrylamide gel electrophoresis (SDS-PAGE). However, the precursors are very short-lived which complicates their detection. In eukaryotes this was overcome by using cell-free translation systems which, in the absence of the membranes that contained the processing enzyme, manufactured precursor proteins. For example, the first study of this kind used mRNA from myeloma cells which was translated in an *in vitro* reticulocyte lysate system into the precursor form of the light chain of immuno-globulin G. This molecule contained an additional peptide of molecular mass 3000. When membranes were added to the translation system, however, the precursor form was converted to mature IgG. This initial work by Milstein and his colleagues in 1972 was rapidly exploited in other eukaryotic systems since it was

5

Extracellular Enzymes

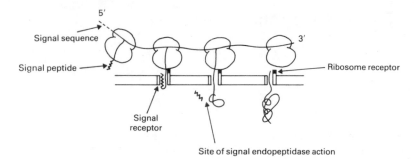

Fig. 1 Schematic diagram of the signal hypothesis for the transport of proteins across membranes (after Blobel *et al.*, 1979). The signal sequence is translated into a signal peptide that recruits one or more receptor proteins in the membrane to form a pore. Similarly, the ribosome binds to a receptor protein. The nascent polypeptide chain is transferred through this pore and the signal sequence is removed by the endoproteolytic action of the signal peptidase. On completion of cotranslational transport, the receptor proteins are free to diffuse in the plane of the membrane.

relatively straight-forward to isolate specific mRNA molecules from specialized eukaryotic cells and to translate them in heterologous cell-free systems based on rabbit reticulocytes or wheat germ cells. The list of eukaryotic secreted proteins known to contain a precursor signal sequence is now substantial (40 to 50; Kreil, 1981).

In bacteria, the same approach could not be used because it was not possible to isolate mRNAs for specific proteins. This stems from the absence of specialized secretory cells devoted to single or relatively few proteins which contain ample quantities of specific mRNAs and from the instability of prokaryotic mRNA. Alternative strategies were therefore adopted to demonstrate precursor forms of secreted proteins. In chain-completion experiments, membrane-bound polysomes are separated from cytoplasmic ribosomes by sucrose gradient centrifugation or gel filtration and then incubated in a suitable medium for the *in vitro* synthesis (completion) of the partially-synthesized polypeptide chains. The products are then identified serologically and characterized by SDS-PAGE. In some instances, coupled transcription and translation systems using specialized transducing phage or cloned DNA templates have also been used to generate precursor forms of secreted proteins. Secondly, *in vivo* procedures have been successful for the study of outer membrane proteins in *E. coli*. Various chemicals that partially disrupt the envelope structure of *E. coli* allow protein synthesis to continue but inhibit the processing of exported proteins. Thus cells treated with toluene or phenethyl alcohol accumulate precursor forms of lipoprotein which can be extracted from the total envelope proteins by precipitation with antilipoprotein antiserum and characterized by SDS-PAGE. Finally, in several instances the presence of a signal sequence has been inferred by comparison of the DNA sequence of the gene with the amino acid sequence of the mature protein.

Using such techniques, many bacterial secreted proteins have been shown to be synthesized as a larger precursor form of the mature protein. In *E. coli* these include some phage-coded, cytoplasmic membrane proteins (phage M13 major and minor coat proteins), several periplasmic proteins including alkaline phosphatase, various binding proteins involved in the transport of small molecules and the plasmid-coded TEM β-lactamase, and some outer membrane proteins including *lamB* protein and lipoprotein. In *Bacillus*, α-amylase and β-lactamase of *B. amyloliquefaciens* and *B. lichenformis*, respectively, are synthesized in larger precursor form.

Structure and function of signal sequences The primary structures of many eukaryotic and prokaryotic signal sequences have now been established by protein or DNA sequencing. They vary considerably in size (15 to 32 amino acids) and superficially there is little homology. However, when the distribution of hydrophobic and polar amino acid residues is compared, it is apparent that all signal sequences contain two particular domains; an NH_2-terminal hydrophilic segment followed by a central core of predominantly hydrophobic amino acids (Figure 2). The hydrophilic region is positively charged and accounts for the variable length of signal sequences. It is generally between one and seven amino acids long and its exact structure does not appear to be too critical since *E. coli* exports rat preproinsulin even when the signal sequence is preceded by 18 additional amino acids at the NH_2-terminus.

The central, hydrophobic core of the signal sequence plays a crucial part in secretion. This section is about 55 Å long (15 to 19 amino acids) and terminates in an amino acid before the cleavage site which has a short side chain (for example alanine or glycine). The hydrophobic region usually contains a proline or glycine residue in the centre (about position 17). Using rules for the prediction of protein secondary structures, it has been determined that these peptides most probably exist in highly ordered conformations and it appears that the ability of the hydrophobic core to form two areas of α-helix split by the polar residue around position 17 is critical for protein export. Indeed mutations of the *lamB* signal sequence that would be expected to destroy these α-helices do inhibit export, and suppressor mutations that should re-establish the secondary structure, allow export.

Several models have been proposed to relate the structure of the signal peptide to its function. Inouye and Halegoua (1980) suggested the loop model in which the basic hydrophilic domain of the signal sequence, as it emerges from the ribosome, attaches to the negatively-charged inner surface of the cytoplasmic membrane by ionic interaction. As the protein is translated, the hydrophobic core of the signal sequence is progressively inserted into the membrane until the cleavage site emerges as a loop on the outside of the membrane. Signal peptidase then cleaves the signal sequence from the nascent, polypeptide chain allowing the protein to be cotranslationally secreted (Figure 3). According to this model, if the charge at the

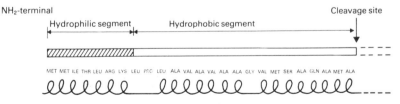

Fig. 2 Structure of the signal peptide of the *lamB* protein of *Escherichia coli*. Signal sequences typically display a hydrophilic NH_2-terminal section of variable length and an internal hydrophobic region of 15 to 19 amino acids. Areas of predicted α-helix (loops) or random coils (straight line) are indicated.

Signal peptidase

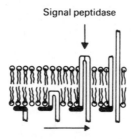

Fig. 3 Loop model for the translocation of secretory proteins across membranes (from Inouye & Halegoua, 1980). The solid portion represents the basic region of the NH-terminal end which attaches to the inner surface of the membrane. The following blank portion represents the hydrophobic region which is progressively inserted into the membrane. When the cleavage site emerges on the outside of the membrane, it is hydrolysed by the signal peptidase (arrow) allowing the protein to be secreted through the membrane. (By permission of CRC Press, Florida).

NH_2-terminus of the signal peptide was negative instead of positive, initiation of export would not occur and the protein would accumulate in the cytoplasm. Such mutants with net negative charges have been prepared by *in vitro* mutagenesis and behave as predicted, but it seems that this model may be a simplification since recent evidence suggests that there may after all be an export machinery in the membrane which is involved in protein translocation.

Cotranslational secretion   The early influential studies of protein transport used specialized secretory cells of animal origin (mainly pancreas and liver cells) in which electron microscopy disclosed two distinct populations of ribosomes; some were apparently attached to the inner surface of the endoplasmic reticulum, while others were free in the cytoplasm. Palade observed a parallel between the abundance of membrane bound ribosomes and the secretion of proteins and suggested that proteins were transported across the membrane as they were synthesized by these ribosomes in a process termed cotranslational secretion. The soluble ribosomes would be responsible for cytoplasmic protein synthesis. Although it was rapidly established that cotranslational secretion was the predominant mode of protein export, it should not be inferred that this is the only process. Post-translational secretion (transfer across the membrane of a completed protein) occurs in both eukaryotic and prokaryotic systems.

It has not been possible to demonstrate the two populations of ribosomes in thin sections of bacteria because of their dense packing in the cytoplasm and the absence of a membrane system analogous to the endoplasmic reticulum. Hence, although membrane fragments in bacterial lysates have long been observed to contain ribosomes, it has not been certain if the attachment was functional or due to artificial association. In the early 1970s, a functional attachment was suggested by the finding that the membrane associated polysomes of *E. coli* produced more secreted protein (alkaline phosphatase) *in vitro* than did the cytoplasmic ribosomes. Improved methods of separating the two populations of ribosomes using sucrose gradient centrifugation or gel filtration achieved more definitive results, and it has since been demonstrated that α-amylase in *B. subtilis* and *B. licheniformis* and numerous periplasmic proteins in *E. coli* are synthesized exclusively by membrane associated ribosomes while the cytoplasmic elongation factor EFTu is made only on soluble ribosomes. It would seem, therefore, that most secreted proteins are transferred across the membrane cotranslationally from membrane bound ribosomes. Nevertheless, these findings could be interpreted as the elongating chain folding against the membrane surface with subsequent engulfment by the membrane after release of the protein from the ribosome.

Direct evidence for cotranslational secretion Cotranslational secretion of a protein could be unambiguously demonstrated if the end of the growing chain protruding from the membrane could be labelled, while the other end remained attached to a ribosome as peptidyl-tRNA. To achieve this, *E. coli* was chilled and treated with chloramphenicol which stabilizes the polysomes. The cells were converted to spheroplasts and labelled with ^{35}S-acetylmethionyl methylphosphate sulphone which reacts with free amino groups of proteins but does not penetrate membranes. The spheroplasts were then washed, osmotically lysed and the membrane associated ribosomes purified by sucrose gradient centrifugation. After removal of the membrane with detergents, a substantial amount of the initial label remained in the polysomes (Figure 4). Moreover the label was presumably attached via the nascent polypeptide chain, since it was released by chain completion *in vitro*. Amongst the translation products, the secreted protein alkaline phosphatase could be identified. These experiments have since been refined and used to establish cotranslational secretion of α-amylase in *B. subtilis*, β-lactamase in *B. licheniformis* and toxin in *Corynebacterium diphtheriae* (Davis & Tai, 1980).

Post-translational secretion Several proteins are incorporated into or transported across membranes after they have been synthesized. This has been demonstrated for proteins located in chloroplasts and mitochondria, which are made on cytoplasmic ribosomes and subsequently imported into the organelle. In bacteria there are also several examples of post-translational secretion. Subunit A of cholera toxin is synthesized *in vitro* by soluble and not by membrane bound

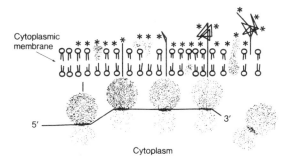

Cytoplasmic membrane

5' 3'

Cytoplasm

Fig. 4 Scheme for extracellular labelling of secreted proteins. The nascent polypeptide chain emerges from the membrane and is labelled with [^{35}S] acetyl methionyl methylphosphate sulphone (asterisks). Subsequent demonstration of label attached to polysomes indicates cotranslational secretion. (From Smith *et al.*, 1977.)

ribosomes and the toxin can be detected immunologically in cell-free extracts of *V. cholerae*. Moreover, in some *Bacillus* strains, α-glucosidase may accumulate in the cytoplasm before being secreted into the environment.

The signal hypothesis can be modified to accommodate post-translational secretion as shown in Figure 5. It is proposed that the protein to be translocated has a signal sequence which remains exposed and interacts with receptors in the membrane to form a pore. Passage through the membrane would be accompanied by unfolding during transfer with subsequent re-folding.

An alternative scheme is the 'membrane trigger' hypothesis. Again a signal sequence is involved, but the function of this peptide is to promote the folding of the protein in such a way that it interacts with the membrane and export is triggered into and across the membrane. Thus no specific export machinery is required (Figure 5). Once on the outside of the membrane, removal of the signal peptide would make the process irreversible. This model is supported by the demonstration that phage M13 coat protein correctly inserts into liposomes comprising nothing but phospholipid and purified processing enzyme.

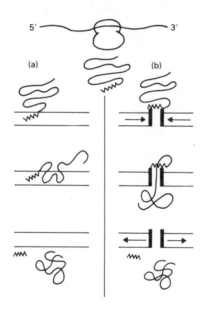

Fig. 5 Models for post-translational transport of proteins across membranes. (a): membrane trigger hypothesis (after Wickner, 1979) and (b): the modified signal hypothesis (after Blobel *et al.*, 1979). In both schemes, a soluble ribosome produces a precursor protein bearing a signal sequence (zig-zag line). In (a), the sequence modifies the folding of the polypeptide such that a configuration with hydrophobic regions associatd with the membrane is formed. The protein may then be released from the membrane possibly with endoproteolytic removal of the signal sequence. In (b), the signal sequence of the completed polypeptide chain recruits receptor proteins in the membrane to form a pore. The protein unfolds and translocation proceeds, followed by removal of the signal sequence.

Processing the precursor   In the original signal hypothesis, the precursor protein is processed by removal of the signal peptide during or immediately after the synthesis of the protein. Recent studies in *E. coli* have provided insights into the stages at which processing occurs. In these experiments, proteins are pulse-labelled and immunoprecipitated to obtain a particular protein. When this precipitate is analysed by SDS-PAGE, an array of molecules is obtained comprising precursor and mature species and incomplete peptides. Since precursor is observed, it indicates that at least some post-translational processing occurs. After limited *in situ* proteolysis of the polypeptides in the gel, the digestion products can be electrophoresed in a second dimension and peptides characteristic of the NH_2-terminus of both precursor and mature forms of the protein can be obtained. It is therefore possible to demonstrate that among the incomplete nascent chains, some still have their signal sequence attached whereas others have been processed and contain an NH_2-terminus characteristic of the mature form (Josefsson & Randall, 1981). In this way, it has been established that the maltose and arabinose binding proteins and alkaline phosphatase of *E. coli* are processed both cotranslationally and post-translationally, while others are processed either during or after translation. For all these proteins, processing is a relatively late event occurring after the protein has been elongated to at least 80% of its final length.

Comparison of the cleavage sites in different precursor proteins reveals little specificity; the peptide is hydrolysed between an amino acid residue with a short side chain (generally glycine or alanine) and the adjacent residue (Figure 2). In eukaryotes, the signal peptidase activity is located in the membrane of the endoplasmic reticulum. An *E. coli* processing enzyme has been purified from both the cytoplasmic and outer membranes in which it is present in roughly equal amounts. This is the first example of such a dual distribution of a membrane protein in *E. coli*.

Genetic studies

The effectiveness of a combined genetic and biochemical approach to the analysis of a process such as protein secretion has been ably demonstrated in *E. coli*. Such studies initially focussed on two questions. (1) If a cytoplasmic protein is provided with an NH_2-terminal sequence from an exported protein, is this sufficient to promote secretion. (2) Do mutations in the signal sequence of secreted proteins alter the final locations of these proteins? However, once these initial studies had been completed and the mutants had been obtained, analysis of their phenotypes led to more rapid isolation of different mutations affecting the secretion process. This approach is now providing evidence of the molecular nature of the secretion process itself.

Protein fusions   DNA can be transposed in *E. coli* and two genes brought into juxtaposition by classical genetic techniques. Such gene fusions code for hybrid proteins and in this way the NH_2-terminal portion of a secreted protein can be fused to a cytoplasmic protein. The genetic manipulations involved are outside the scope of this book and have been fully described by Silhavy *et al.* (1979).

Strains of *E. coli* have been developed in which the *lacZ* (β-galactosidase) gene has been fused to various loci in the *mal* (maltose) utilization operon; *malF*, which codes for the maltose transport protein located in the cytoplasmic membrane;

Extracellular Enzymes

malE, which codes for the periplasmic maltose-binding protein; and *lamB* which codes for an outer membrane protein that is essential for maltodextrin transport and also serves as the receptor for phage λ. Since the fused genes retain a substantial portion of *lacZ*, the hybrid proteins display β-galactosidase activity which can be located in the cell by fractionation procedures. Hybrid proteins comprising the maltose transport protein (*malF*) were cytoplasmic if the proportion of *malF* DNA was small. However, one class of fusions contained a substantial portion of *malF* and the hybrid protein was located in the cytoplasmic membrane which is where the *malF* product normally resides. Similarly, certain *lamB-lacZ* fusion-bearing strains that contained most of the *lamB* gene exported the hybrid protein to the outer membrane, the normal location of the *lamB* product. However, with *malE-lacZ* fusions, β-galactosidase activity was confined to the cytoplasm if a small portion of *malE* was involved, and located in the cytoplasmic membrane if a large portion was present. Similar results have been obtained with another periplasmic protein (alkaline phosphatase), and in no instance has any hybrid protein containing β-galactosidase activity been secreted into the periplasm. A possible explanation for these findings is that export is initiated by these hybrid proteins but then aborted because certain amino acid sequences within β-galactosidase are incompatible with passage through the membrane. Thus the *malF-lacZ* products remain in the cytoplasmic membrane and the *lamB-lacZ* products travel via the interconnecting bridges (Bayer's patches) from the cytoplasmic membrane to the outer membrane, but neither is released from the membrane.

Certain *malE-lacZ* fusions confer maltose sensitivity on the host cell. Upon addition of maltose to the culture, the hybrid protein is induced and export is initiated but, because the protein cannot be released from the membrane, it blocks the export of other proteins resulting in cell death. Mutations could therefore be readily selected for suppression of the maltose sensitive phenotype without loss of maltose inducible β-galactosidase activity, by selecting for β-galactosidase activity in the presence of maltose. Such mutants had altered signal sequences such that the hybrid β-galactosidase was located in the cytoplasm rather than the membrane. DNA sequence analysis of these mutant genes revealed deletions or point mutations that altered the structure of the hydrophobic core of the signal sequence. All export-defective mutations of the signal sequence result in the presence of charged amino acids in this region, although recent experiments suggest that the position of this charge is also important. Export will be blocked if the formation of the α-helical structures (Figure 2) is inhibited. This work has been reviewed recently (Michaelis & Beckwith, 1982).

Analysis of the carboxyl-terminus of secreted proteins The signal hypothesis proposes that the NH_2-terminus of a secreted protein initiates export and the remaining portion of the protein passes through the membrane in a passive manner. In the case of the intrinsic membrane proteins, a 'stop secretion' sequence would halt secretion. Analysis of the periplasmic protein β-lactamase using chain-terminating mutants that result in loss of specific amounts of the COOH-terminus has shown that this protein is exported correctly but not released into the periplasm following modification of the COOH-terminus. This end of the protein is therefore not important in export, but is presumably involved in the release or solubilization of the enzyme.

Analysis of the secretion apparatus The exact nature of any molecular export machinery in the ribosome or membrane of *E. coli* remains unclear, but genetic studies indicate an important role both for ribosomes and for membrane components that may resemble the eukaryotic translocation apparatus. Mutants of *E. coli* defective in the secretory apparatus have been isolated by selecting for suppression of the export-defective mutants described above. For example, mutations in the *lamB* signal sequence restrict the protein to the cytoplasm and the organism is unable to transport and consequently grow on maltodextrins. Suppressor mutations that permit growth on this carbon source restore *lamB* product to the outer membrane. These mutations map in three loci, *prl* (for protein localization) *A*, *B* and *C* of which *prlA* mutants have been the most extensively studied. The suppression is not specific: signal sequence defective mutants of *malE* and *phoA*, in which the respective proteins are not exported to the periplasm but remain in the cytoplasm, behave as wild type and localize the maltose-binding protein and alkaline phosphatase correctly in *prlA* hosts. The *prlA* gene maps within an operon coding for ribosomal proteins and it is likely that it codes for a ribosome component.

An alternative approach to the identification of components involved in secretion is to isolate mutants impaired in the export of cell envelope proteins. A selection for such mutants was developed when it was discovered that β-galactosidase displayed negligible activity when it was located, as a *malE-lacZ* hybrid product, in the cytoplasmic membrane. Such strains are unable to grow on lactose, possibly because the enzyme is unable to assume its tetrameric structure in this location. By selecting for growth on lactose, mutants in which the hybrid protein is cytoplasmic were obtained. Some of these strains had a defect in the signal sequence of the *malE* portion of the gene, thus inhibiting export, but others had mutations in other parts of the chromosome. These *sec* (secretion defective) mutants are of two classes, *secA* and *secB*, of which the former have been characterized. The *secA* gene codes for a membrane protein that seems to be involved in export, since *secA* mutants accumulate precursors of a variety of periplasmic and outer membrane proteins (Michaelis & Beckwith, 1982).

Secretion in eukaryotic cells

The original signal hypothesis was based on findings from eukaryotic cells and proposed that the signal peptide, as it emerged from the ribosome, interacted with receptor proteins in the membrane to form a tunnel that was subsequently stabilized by ribosome binding. Early biochemical studies identified trans-membrane glycoproteins that associate with membrane bound ribosomes. Evidence that these or similar proteins are involved in protein export has been provided by showing that microsomal membranes translocate nascent proteins *in vitro*, but fail to do so after salt-washing or trypsin treatment which removes or inactivates respectively the membrane proteins.

However, there is a timing problem in the translocation process. Proteins are being synthesized very rapidly *in vivo* and, the signal sequence, as it emerges from the ribosome, would have to locate the membrane within a few seconds, since further chain elongation would allow the formation of a secondary structure and probably block the signal sequence. Such a problem could be circumvented if there were a temporary halt to translation until the ribosome had located the membrane. This idea has been confirmed by the isolation and characterization of the signal recognition particle. Comprising six polypeptide chains and a small (7S) RNA molecule, this complex blocks cell-free translation of secreted proteins as soon as the signal peptide emerges from the ribosome. Thus the ribosome-mRNA complex has an indefinite period to find the membrane. Concomitant with characterization of the recognition complex, a 'docking' protein was identified in the membrane of the endoplasmic reticulum. *In vitro,* this protein abolishes the inhibition of translation caused by the recognition complex and allows cotranslational secretion to continue (Figure 6). *In vivo* it is proposed that the ribosome-signal recognition particle complex forms as soon as the signal peptide emerges from the large ribosomal subunit. Translation is temporarily halted but, when the complex associates with the docking protein in the membrane, translation proceeds and secretion of the nascent protein ensues (Meyer *et al.,* 1982).

Secretory pathway in fungal cells

The added complexity of the eukaryotic cell complicates protein secretion in fungi and the intermediates in the secretory pathway have been difficult to establish because export is a rapid process. Addition of cycloheximide to a derepressed culture of *S. cerevisiae* inhibits invertase secretion within five minutes, indicating that inhibition of translation and completion of export occurs within this period. The secretory pathway has therefore been analysed using temperature sensitive mutants that are deficient in secretion. These mutants accumulate precursors of secreted proteins at specific stages in the pathway upon incubation of the cells at the non-permissive temperature. The secretory pathway appears to be similar to that in higher eukaryotes and involves a series of membrane-bounded structures which mediate the transfer of the protein from the site of synthesis at the rough endoplasmic reticulum to the site of secretion at the plasma membrane. In effect, the protein is discharged into the cisternal space of the rough endoplasmic reticulum where the initial steps of glycosylation occur. In an energy dependent step involving at least nine gene products, the protein is transferred to a Golgi-like structure where further glycosylation occurs. Two or more gene products and energy are then required to package the glycoprotein into secretory vesicles that

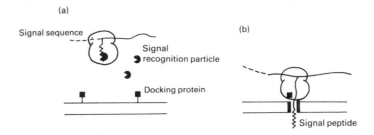

Fig. 6 Initial events in the vectorial transfer of proteins in the eukaryotic cell. (a) The signal recognition particle binds to the ribosome which bears a signal peptide. Translation is stopped. (b) The ribosome interacts with the docking protein, the block is removed and translation proceeds coupled with secretion of the protein.

migrate and fuse with the plasma membrane in the area of the growing bud, in a third energy dependent step which requires at least 10 additional gene products. Cytoplasmic vesicles and Golgi apparatus have been noted in the subapical zone of some filamentous fungi, so it is possible that this scheme may also apply to some moulds. However, an ultrastructural study of *Trichoderma viride* revealed no recognizable Golgi apparatus, suggesting that secretory pathways may vary amongst the fungi. Nevertheless, the genetic approach adopted by Scheckman and his colleagues for analysis of the secretory pathway in *S. cerevisiae* should be useful for the study of these other fungi.

Summary

In Gram positive bacteria and fungi, some enzymes are transported through the cytoplasmic membrane, diffuse through the cell wall and accumulate in the environment. In Gram negative bacteria, the cell wall comprises two membranes separated by the periplasmic zone in which many enzymes are located. A key feature of all these enzymes is that they have crossed the cytoplasmic membrane and are therefore extracellular. Proteins that traverse membranes typically possess an NH_2-termini peptide extension of about 25 amino acid residues; the signal peptide. As the signal peptide emerges from the ribosome, it interacts with the inner surface of the cytoplasmic membrane in bacteria or the endoplasmic reticulum in eukaryotes. As the polypeptide is elongated it passes through the membrane (cotranslational secretion) and a processing enzyme (signal peptidase) removes the signal peptide on the outer side. This allows the protein to assume its normal configuration on the outside of the membrane. Genetic studies have shown that there is a secretory machinery in the *E. coli* envelope that appears to transport the protein actively. The secretion of proteins is remarkably similar in eukaryotic and prokaryotic cells to the extent that eukaryotic signal sequences are recognized and processed by prokaryotic cells and *vice versa*. However, the additional complexity of the eukaryotic cell requires that a series of membrane bounded vacuoles are needed to transport the enzyme from the endoplasmic reticulum to the outside of the cell. Cotranslational secretion is not universal, however, and some proteins are incorporated into, or transported through membranes, after they have been synthesized (post-translational secretion), again through the involvement of a signal peptide.

Genes for cytoplasmic enzymes have been fused to those for exported enzymes in *E. coli* and the resultant hybrids comprise cytoplasmic proteins with a signal sequence. Such proteins are incorporated into the membrane, and may be released into the periplasm of *E. coli* or secreted into the medium by *B. subtilis*. Such systems hold great promise for industrial production of proteins from microorganisms.

References

BLOBEL, G., WALTER, P., CHANG, C. N., GOLDMAN, B. M., ERICKSON, A. H. and LINGAPPA, V. R. (1979). Translocation of proteins across membranes: the signal hypothesis and beyond. *Symposium of the Society for Experimental Biology*, 33, 9–36.

Extracellular Enzymes

DAVIS, B. D. and TAI, P.-C. (1980). The mechanism of protein secretion across membranes. *Nature* 283, 433–438.

EMR, S. R. and SILHAVY, T. J. (1982). Molecular components of the signal sequence that function in the initiation of protein export. *Journal of Cell Biology* 95, 689–96.

INOUYE, M. and HALEGOUA, S. (1980. Secretion and membrane localization of proteins in *Escherichia coli. CRC Critical Reviews in Biochemistry* 7; 339–71.

JOSEFSSON, L. G. and RANDALL, L. L. (1981). Different exported proteins in *E. coli* show differences in the temporal mode of processing *in vivo. Cell* 25; 151–7.

KREIL, G. (1981). Transfer of proteins across membranes. *Annual Review of Biochemistry* 50; 317–48.

MEYER, D. I., KRAUSE, E. and DOBBERSTEIN, B. (1982). Secretory protein translocation across membranes – the role of the 'docking protein'. *Nature* 297; 647–50.

MICHAELIS, S. and BECKWITH, J. (1982). Mechanism of incorporation of cell envelope proteins. *Annual Review of Microbiology* 36; 435–65.

NOVICK, P., FERRO, F. and SCHEKMAN, R. (1981). Order of events in the yeast secretory pathway. *Cell* 30; 439–48.

SILHAVY, T. J., BASSFORD, Jr., P. S. and BECKWITH, J. (1979). A genetic approach to the study of protein localization in *Escherichia coli.* In: *Bacterial Outer Membranes,* pp. 203–54. Edited by M. Inouye. John Wiley & Sons, New York.

SMITH, W. P., TAI, P.-C., THOMPSON, R. C. and DAVIS, B. D. (1977). Extracellular labelling of nascent polypeptides traversing the membrane of *Escherichia coli. Proceedings of the National Academy of Sciences USA* 74; 2830–4.

WICKNER, W. (1979). The assembly of proteins into biological membranes: the membrane trigger hypothesis. *Annual Review of Biochemistry* 48; 23–45.

3 Regulation of extracellular enzyme synthesis

Early studies of extracellular enzyme synthesis involved industrially important enzymes, and were simply attempts to maximise enzyme yields by variation of the growth conditions. These experiments provided a wealth of information on the induction and repression of extracellular enzymes, but nothing on the molecular systems involved. This situation exists to the present, largely because the organisms used for the commercial production of extracellular enzymes lack any exploitable genetic system. With the introduction and rapid application of recombinant DNA techniques to a variety of microorganisms, we now have the ability to purify DNA fragments from virtually any source and place them in specific genetic backgrounds; commonly *Escherichia coli*, *Bacillus subtilis* or *Saccharomyces cerevisiae*. This is providing an opportunity for genetic analysis outside *E. coli* and an insight into genetic regulation in other microorganisms; consequently our perception of the regulation of extracellular enzyme synthesis should increase dramatically within the next few years. However, our current understanding of this area is limited. Therefore, in this chapter I shall describe the effect of growth conditions on the regulation of extracellular enzyme synthesis in various microorganisms, but restrict discussion of the molecular mechanisms involved largely to *E. coli*.

Environmental control of extracellular enzymes

Enzyme induction   If the rate of enzyme synthesis is constant irrespective of the presence of substrate in the environment, the enzyme is described as constitutive. Several extracellular enzymes are constitutive, including the industrially important amylases and proteases of *Bacillus amyloliquefaciens* and *Bacillus licheniformis*, but it is more common that enzyme synthesis is inducible. Inducible enzymes are synthesized at a low basal rate in the absence of substrate. When the substrate, or a derivative thereof, is present in the medium there is a dramatic increase in the rate of synthesis of the particular enzyme(s) for its catabolism. Synthesis continues at this amplified rate until the inducer is removed; it then returns to the basal rate.

The efficient induction of extracellular enzymes is slightly more complicated than that for intracellular enzymes. Firstly, since many of the substrates are too large to enter the cell, the organism must have some means of detecting the macromolecule in the environment. Secondly, for the enzyme to be secreted at appropriate rates, the organism must be able to monitor the activity of the enzyme outside the cell in what may be a very harsh environment. It has been proposed that these problems are overcome by a low constitutive rate of enzyme secretion. In the presence of substrate, low molecular weight products accumulate which enter the cell and effect induction. If the external environment inhibits the enzyme (due for example to an unsuitable pH or an extreme of temperature), induction would not occur since the products would not be generated. However, to limit secretion should an excess of products accumulate in a conducive environment, synthesis of the enzyme is inhibited by some form of end-product repression. In this way, the organism gains a measure of control over enzyme secretion and prevents wasteful synthesis. Many physiological observations support this scheme in various bacteria and fungi. The basal rate of synthesis of inducible extracellular enzymes varies from a barely detectable rate which is increased several thousand-fold on induction, to a relatively high rate that is best described as partially constitutive and which doubles or trebles on induction.

In general, pectinases are inducible and the pectic acid lyase of *Erwinia carotovora* has been studied in detail. The principal product of the cleavage of pectic acid by the enzyme is an unsaturated digalacturonic acid residue, which specifically induces the enzyme. However, too high a concentration of this inducer represses pectic acid lyase synthesis, presumably by catabolite repression since it can be reversed by the addition of cyclic adenosine $3',5'$-monophosphate (cAMP). Similarly the α-amylases of some strains of *Bacillus stearothermophilus* and *B. licheniformis* hydrolyse starch predominantly to maltotetraose and malto-pentaose respectively, and these molecules induce enzyme synthesis. These systems are not as specific as the pectate lyase of *E. carotovora*, however, and maltooligosaccharides ranging from G3 to G7 will effect the induction. Both these enzymes are susceptible to catabolite repression and accumulation of glucose in the environment totally inhibits synthesis.

Similar systems operate in eukaryotes. The cellulase complex of enzymes synthesized by *Trichoderma reesei* is induced by cellulose but most low molecular weight products of cellulose degradation are poor inducers. It is thought that such molecules may be rapidly degraded to glucose, and thus cause catabolite repression, since sophorose, a $1,2$-β disaccharide of glucose is less rapidly hydrolysed and induces the cellulase complex strongly. The close relationship between induction and catabolite repression has caused considerable confusion. For most extracellular enzymes, non-metabolizable ('gratuitous') inducers have not been found. Consequently, when an 'inducer' is added to a culture it is metabolized and the growth of the culture is modified. Thus the appearance of an enzyme may be the result of induction or a change in growth rate which releases the organism from catabolite repression. Thus careful and rigorously defined experiments are required to provide unequivocal evidence for enzyme induction, preferably involving the use of gratuitous inducers.

Enzyme repression As mentioned above, the product of an enzyme's activity may inhibit its synthesis. This can take two forms; if the repression is specific for the particular enzyme it is often referred to as end-product repression. Alternatively a range of enzymes involved in a certain aspect of metabolism, principally nitrogen, carbon or phosphorous may be coordinately repressed. Repression of enzymes concerned with nitrogen metabolism, particularly those enzymes that supply ammonia or glutamate, by a plentiful supply of amino acids or ammonia is termed nitrogen regulation or nitrogen catabolite repression. Similarly, the repression of enzymes involved in the catabolism of exotic carbon sources by rapid growth on a readily utilized carbon source, is carbon catabolite repression or simply catabolite repression. Moreover, high phosphate concentrations often repress enzymes involved in phosphorus metabolism such as alkaline phosphatase, deoxyribonuclease and ribonuclease.

The synthesis of extracellular proteases is commonly controlled by repression.

Protease synthesis in bacilli is typically constitutive but subject to repression. It is not certain if this is end-product repression or nitrogen regulation. A variety of amino acids and peptides repress protease synthesis in the different species; for example glutamate and aspartate repress *B. subtilis* protease synthesis strongly but have little effect in *B. megaterium* in which isoleucine and threonine repress effectively. In continuous cultures of *B. licheniformis*, protease secretion only occurs under nitrogen limiting conditions, providing evidence that a nitrogen-rich environment will repress protease synthesis through nitrogen regulation. In other microorganisms including vibrios and *Neurospora crassa*, protease synthesis is induced by low concentrations of amino acids and repressed by high concentrations, or by ammonium ions. Again nitrogen regulation is probably involved.

Carbon catabolite repression is the permanent repression of inducible or constitutive enzyme synthesis that occurs in most microorganisms growing on a readily utilized carbon source such as glucose or glycerol. Furthermore, if such a carbon source is added to a culture growing slowly on a poor carbon source, there is a severe transient repression of catabolite sensitive operons, the expression of which subsequently adopts the catabolite repressed rate. The synthesis of virtually all extracellular enzymes in most microorganisms, including proteases, nucleic acid hydrolysing enzymes and carbohydrases, is controlled by catabolite repression. In this way, a microorganism can avoid the wasteful synthesis of inducible or constitutive enzymes for the utilization of relatively poor carbon sources in the environment. However, although the phenomenon is similar throughout the microbial world, it is important to note that the molecular mechanisms through which it operates appear to vary considerably, and will be described in some detail later.

Patterns of extracellular enzyme synthesis Extracellular enzyme synthesis generally conforms to one of two patterns. Synthesis and secretion may accompany growth and decline as the cells enter stationary phase. Alternatively, enzymes may be synthesized at a minimal rate during exponential growth but accumulate in large amounts during stationary phase. These patterns represent the extremes and in some cases may merge. Moreover, the production of the enzyme may depend on growth conditions; neutral protease synthesis in *B. megaterium* accompanies growth in minimal medium but only appears during stationary phase when complex nitrogen sources are used.

These patterns of extracellular enzyme synthesis have important implications for commercial enzyme production. If a protein is synthesized only during stationary phase it might be expected that continuous cultures maintained in constant exponential growth in a chemostat would make very little, if any, of the product. This is indeed the case, and α-amylase production by chemostat cultures of *B. amyloliquefaciens* is poor and variable. However, for those enzymes that are secreted during exponential phase, the chemostat provides an ideal growth environment and steady state levels of extracellular protein can be readily achieved.

In conclusion, to understand completely the regulation of extracellular enzyme synthesis, we must be able to account for induction and repression, including nitrogen and carbon catabolite repression, and to explain how the synthesis of some enzymes is derepressed either during exponential growth or during stationary phase. These aspects of the regulation of extracellular enzyme synthesis will be examined in the latter part of this chapter.

Protein synthesis

In prokaryotes, proteins are synthesized in two stages. One strand of the DNA is first transcribed by RNA polymerase into a complementary copy of RNA, the messenger (m)RNA. Translation of this mRNA ensues as soon as the ribosome binding site is available. Thus the two stages of protein synthesis are generally coupled events since the transcript is both elongated and translated in the same (5′ → 3′) direction. A number of ribosomes may translate a mRNA at any one time; such aggregates are termed polyribosomes or polysomes. The synthesis of secreted proteins is probably an exception to this rule since the polysomes translate the message while it is attached to the inside surface of the cytoplasmic membrane. Transport of mRNA from the transcription site on the DNA to the membrane is therefore necessary and it would appear that transcription and translation may not be so tightly coupled for these proteins (see Chapter 2).

At this point it will be useful to compare the specificity of protein synthesis in Gram negative and Gram positive bacteria. With the development of *B. subtilis* as a host for gene cloning, it soon became apparent that foreign genes such as those from *E. coli* were not expressed in *B. subtilis* although *B. subtilis* genes were expressed in *E. coli*. There are considerable differences in the RNA polymerases of these two bacteria. RNA polymerase comprises a core enzyme consisting of four subunits that effects transcription. In *E. coli*, core enzyme combines with a polypeptide of molecules mass 95 000, the sigma (σ) factor that directs transcription to the promoter. Without σ, transcription initiates at a variety of non-specific sites. A major difference between Gram positive and Gram negative bacteria is the multitude of polypeptides associated with RNA polymerase in the former compared to the single RNA polymerase in the latter. Although many of these forms are thought to play a role in sporulation, this is also a feature of other non-sporulating Gram positive bacteria. The major form of RNA polymerase in *B. subtilis* comprises core enzyme associated with a σ factor of molecular mass 55 000 but during sporulation, core enzyme acquires a variety of new polypeptides of which at least one is involved in the initiation of the sporulation process.

Despite the gross differences in RNA polymerase structure between *B. subtilis* and *E. coli*, promoters for genes in the two organisms appear to be quite similar. Those promoter sequences from *B. subtilis* that have been characterized share similar 'consensus' sequences with *E. coli* promoters, particularly in the highly conserved Pribnow box and '–35' regions. It is therefore not surprising that no absolute transcription barrier appears to exist between these species. Indeed *B. subtilis* and *E. coli* genes are transcribed *in vitro* by the heterologous RNA polymerases. Nevertheless, several studies have shown that the *B. subtilis* enzyme is considerably less efficient at transcribing *E. coli* templates than *vice versa*. The principal barrier to heterologous gene expression must therefore occur at the translational level. Evidence for this was obtained by Stallup and Rabinowitz when they attempted to obtain a protein synthesizing system from clostridial cells. Clostridial ribosomes showed activity comparable to *E. coli* ribosomes when clostridial mRNA was used as a template but failed to translate an *E. coli* message. Further studies revealed that this was a general feature of Gram positive and Gram negative cells in that ribosomes from Gram negative bacteria will generally translate both types of mRNA but those from Gram positive bacteria will not translate mRNA from Gram negative organisms. Of the several possible molecular mechanisms for this specificity, the requirement for initiation factors appears to be important. Translation of mRNA from Gram positive sources is relatively independent of initiation factors but in Gram negative bacteria these factors are essential. This has been at least partly explained by comparison of the sequences for ribosome binding in the two types of mRNA. The ribosome binding site of an mRNA is thought to consist of a sequence of at least 3 to 9 bases called the Shine-Dalgarno sequence. This sequence is complementary to the 3' end of the 16S rRNA and probably promotes binding by base pairing with that RNA. Almost invariably in *B. subtilis* there is a high proportion of perfect complementarity of the Shine-Dalgarno region of the mRNA with the 3' end of the 16S rRNA. The GGAGG sequence of the mRNA can form four GC pairs with the CCUCC of the 16S rRNA; in *E. coli* complementarity beyond three or four bases is rare. This suggests that the binding strengths of the Gram positive initiation sites may be greater than those of Gram negative bacteria. This may be an absolute requirement in Gram positive bacteria and could explain the lack of expression of many genes when cloned in *B. subtilis* as well as the relative independence of initiation factors. Moreover, it offers the possibility of engineering the expression of genes from Gram negative bacteria in *B subtilis* by providing the correct ribosome binding site.

Gene expression in the eukaryotic cell is considerably more complex than that in bacteria. The primary transcript from the eukaryotic chromosome is not mRNA but a high molecular weight RNA ranging from 2000 to 20 000 nucleotides in length. Several 'processing' steps occur to this primary transcript before the mRNA is transported into the cytoplasm for translation. Segments (introns) from the polynucleotide are removed and the remaining sequences (exons) 'spliced' together to give rise to a translatable message. Although some mRNAs are derived from primay transcripts of similar size to the finished mRNA (particularly in lower eukaryotes) the majority are probably derived from transcripts considerably larger than the mRNA itself. Both ends of the mRNA are modified. A methylated oligonucleotide 'cap' is added to the 5'-end shortly after the start of transcription. This appears to promote mRNA stability and also appears to be necessary for efficient translation. A long stretch of adenine residues [poly (A)] is added post-transcriptionally but before the final splicing step. It has therefore been suggested that eukaryotic genes are subject to regulation at more levels than are bacterial genes. The processing of the nuclear RNA may not be automatic and the selection of sequences for export to the cytoplasm for translation could be an important mode of control. Nevertheless, for those enzymes from eukaryotic microorganisms in which the applied microbiologist is interested it would seem that regulation of transcription is the principal control point.

Regulation of transcription

The control of protein synthesis in both bacteria and eukaryotes is exerted primarily at transcription. This has obvious advantages to the cell, since it is wasteful to produce transcripts that may not be translated. Experimental evidence for regulation of transcription requires the demonstration that the rate of mRNA synthesis is modulated upon induction or repression of an enzyme. Such evidence has been obtained by labelling mRNA in cells and hybridizing it to DNA sequences of specific genes. Increased hybridization when the enzyme is induced shows the increase in transcription. In this way, the rate of transcription of the lactose-utilizing enzymes of *E. coli* has been shown to vary according to the level of induction or repression. Similar experiments have confirmed transcription as the major control point of protein synthesis in eukaryotes although the complexity of the eukaryotic cell offers the possibility of additional control points as mentioned above.

One specific problem with extracellular enzymes has been the technical inability to prepare specific genes for use as hybridization probes for similar studies in *Bacillus* or fungi. This should be remedied shortly with the introduction of gene cloning in *Bacillus* and the availability of genes in plasmid and phage vectors for hybridization studies. In the meantime, indirect evidence using relatively non-specific hybridization assays and transcription and translation inhibitors suggests that the synthesis of extracellular enzymes is largely, but perhaps not entirely, controlled at the level of transcription.

Positive and negative control systems   In bacteria, groups of related structural and regulatory genes comprise operons that are under common control. These operons usually have a single promoter site to which RNA polymerase binds prior to initiating transcription. The expression of a gene, or set of genes, can therefore be controlled by placing a regulatory sequence between the promoter and the structural genes. An operon in which the regulatory locus (operator) allows free passage of RNA polymerase but is progressively closed by binding a specific repressor molecule is under negative control. Conversely, positive control defines an operator that prevents RNA polymerase from transcribing the structural gene but binds a specific activator that encourages passage of the enzyme (Figure 7). These two modes of regulation govern both inducible and repressible enzymes in bacteria.

The genes for both lactose (*lac*) and galactose (*gal*) utilization in *E. coli* are controlled by negative systems. The repressor proteins are synthesized constitutively from their own transcription units and possess two binding sites, one for the operator DNA and one for the inducing molecule. In the absence of inducer, the repressor binds with a high affinity to the operator DNA reducing transcription to a very low level. The operator follows the promoter in the *lac* operon, so the repressor does not interfere with RNA polymerase binding, but it does prevent transcription progressing into the structural genes. The *gal* operator, on the other hand, overlaps with the promoter so that bound repressor will directly interfere with RNA polymerase binding. The inducing molecule, which may be the substrate for, or an intermediate in, the catabolic pathway (e.g. allolactose for the *lac* operon, galactose for the *gal* operon) interacts with the repressor and greatly lowers its affinity for operator DNA, with the result that the repressor is released and transcription ensues.

Typical catabolic operons under positive control in *E. coli* include the maltose and arabinose operons. The latter is of particular interest since the regulatory protein has a dual role. In the absence of inducer, it acts as a repressor and prevents transcription by binding to its operator but, in the presence of arabinose, it assumes its activator configuration in which it binds to an adjacent site from which it enhances transcription. Since the activator activity is the more powerful of the two, this operon is considered to be under positive control.

Positively and negatively controlled operons can be distinguished genetically. Deletions or nonsense mutations in the gene for the regulatory protein in negative systems give rise to no repressor, and therefore constitutive expression of the

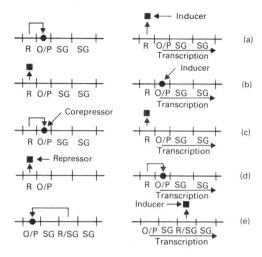

Fig. 7 Control of gene expression. A regulatory gene *R* directs the synthesis of a protein that interacts with the operator site *O* adjacent to the promoter, *P* and controls the transcription of the structural genes, *SG*. (a) Inducible operon under negative control. The regulatory gene codes for a repressor that prevents transcription. In the presence of inducer, the repressor is inactivated and transcription proceeds. (b) Inducible operon under positive control. *R* codes for an activator. Only in the presence of inducer can the activator enhance transcription. (c) Repressible operon under negative control. The repressor (aporepressor) is inactive but converted to an effective repressor by the product of the operon, the co-repressor. (d) Repressible operon under positive control. In the absence of the product of the operon, the activator, coded by the regulatory gene, promotes transcription. In the presence of the product, the activator is inactivated. (e) Inducible operon under autogenous negative control. The *R* gene lies within the operon and may be one of the enzymes of the operon. It acts as in (a) above.

structural gene(s). Conversely, deletion of a gene for an activator protein gives rise to a negative phenotype, since expression cannot be amplified beyond the low, uninduced level. Such genetic studies can be extended using cells that are diploid for the genes in question. For negatively controlled systems, the constitutive phenotype caused by a loss of repressor will be masked by a second and unimpaired copy of the regulatory gene in the same cell, since this functional repressor will be active at both operator sites. Thus the inducible phenotype will dominate over the constitutive phenotype and the cell will behave as wild type (Figure 8). In positively controlled operons on the other hand, a constitutive cell arising from a mutated activator protein will contain an activator that will enhance transcription of the operon whether inducer is present or not. This activator will be effective despite a wild type (inducible) allele in the same cell, and the constitutive phenotype will dominate. Thus for positively controlled systems, the constitutive is dominant over the inducible phenotype.

Returning to extracellular enzymes, there has been virtually no genetic analysis of inducible systems. The only exception is penicillinase (β-lactamase) synthesis in *B. licheniformis*. This enzyme is coded by a single gene, *penP*, which is controlled by a closely linked regulatory gene, *penI*. Constitutive cells have been created through mutations in *penI*, indicating that this locus codes for a repressor protein.

Extracellular Enzymes

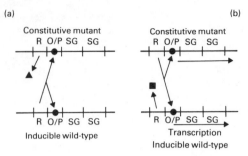

Fig. 8 Dominance relationships of inducible operons in diploid cells: (a) The inducible, negatively controlled operon. The constitutive mutant codes for a faulty repressor (triangle) that fails to bind to operator DNA thus allowing transcription in the absence of inducer. A wild type (inducible) regulatory gene is dominant over the mutant constitutive allele since repressor protein is freely diffusible and will regulate transcription from both chromosomes. (b) Inducible, positively controlled operon. The constitutive mutant codes for a faulty activator that will promote transcription in the absence of inducer. The wild type allele codes for an activator that requires inducer to promote transcription and is therefore recessive to the constitutive genotype.

More recently diploid cells have been prepared by cloning these genes into plasmid vectors. The inducible phenotype was dominant over the constitutive phenotype in these diploids, confirming that *penI* does code for a repressor, but the system is complicated by a second regulatory element the 'effector' or 'anti-repressor' the exact function of which is unknown. Nevertheless, it is likely that extracellular enzyme synthesis in bacteria is controlled by similar systems to those that operate for cytoplasmic catabolic enzymes.

With regard to repressible operons, these are generally responsible for bio-synthetic pathways such as those for amino acid production, since it is advantageous to the cell to inhibit expression of these genes when a plentiful supply of the particular product is in the environment. Both positive and negative control circuits can be devised for such operons (Figure 7) but negative control appears to predominate in those microorganisms that have been examined to date. In these systems, the regulatory gene codes for an apo-repressor that has little affinity for operator DNA and thus the structural genes are expressed. The product of the operon acts as a co-repressor and combines with the apo-repressor to form a functional repressor that binds strongly to the operator locus and inhibits expression of the structural genes. Positively controlled repressible operons use an activator protein that enhances transcription but is inactivated when the product of the operon accumulates. These operons are rare, but can be distinguished from their negatively controlled equivalents since mutations in the activator gene will give rise to a negative phenotype, while mutations in the repressor gene will create a constitutive phenotype in the same way as with inducible operons.

Autogenous control The operons described above are often referred to as 'classical' operons since they conform to the original formulation of the *lac* operon in which the repressor protein is synthesized constitutively from its own promoter. It is possible, however, to construct an operon in which the regulatory gene lies alongside the structural genes of the operon and forms part of the same transcription unit. Alternatively, the product of one of the structural genes may have a dual function as a regulatory protein and as an enzyme. Such operons are autogenous (Figure 7) and can be hypothesized for inducible or repressible, negative or positive systems. Statistical analyses of such operons have led to the conclusion that autogenous control provides unstable or unsatisfactory networks for activator-mediated control but may be advantageous for repressor-mediated control. It can be envisaged that, as the level of induction increases, so does the level of repressor. Consequently, to provide continued expression of the operon, a relatively high and constant presence of inducer is necessary. Since repressor concentration is directly proportional to the level of induction, the system is buffered against rapid fluctuations in the external concentration of inducer.

Autogenously regulated operons can be detected genetically since the deletion of the regulatory gene in the operon should give a negative phenotype for this product but constitutive (in the case of repressor-control) synthesis of the remaining products of the operon. Such evidence should be confirmed by measuring the level of repressor in the cell under different conditions of induction and repression. The nitrite and nitrate reductases of *Aspergillus nidulans* are regulated autogenously. Similarly the histidine utilization operon of *Salmonella typhimurium* is an autogenous negative control circuit. It is not known if any inducible extracellular enzymes are regulated in this way. Detailed descriptions of these various operon configurations have been published by Savageau (1979).

Attenuation   Regulation of gene expression by premature termination of transcription (attenuation) occurs in several amino acid biosynthetic operons as an addition to, or in place of, operator/repressor control (Yanofsky & Kolter, 1982). The attenuator provides repression of gene expression when the respective amino acid is plentiful in the environment. The mRNA for these operons bears a leader sequence at the 5′-end which includes a large proportion of residues which code for the amino acid whose biosynthesis is determined by the operon. The leader can form one of two alternative secondary structures through base pairing; the choice depends on the presence of ribosomes translating the leader, and this in turn is dependent on the availability of charged tRNAs for this amino acid. In the absence of the particular amino acid, charged tRNAs for the amino acid will be scarce, and the ribosomes will stall at those codons in the leader that specify the amino acid. A stem and loop structure will form in the leader sequence by base pairing that allows transcription by RNA polymerase to proceed from the leader sequence into the structural genes. Thus a complete transcript is produced and the genes are expressed. In the presence of the amino acid, the ribosomes will not stall in the leader sequence since the supply of charged tRNAs will be plentiful. The ribosomes will prevent the former secondary structure occurring, but will allow an alternative stem and loop to form. This structure signals the RNA polymerase to terminate transcription prematurely and the structural genes are not expressed. In certain cases, for example the *his* operon, attenuation is the only mechanism of regulation; in others such as the *trp* operon it is additional to repressor-operator control.

Attenuation systems are not solely involved in amino acid biosynthetic operons. It has been suggested that β-lactamase synthesis in *E. coli* may be regulated in a similar way. The relative amount of chromosomally encoded β-lactamase of *E. coli* increases with growth rate. It is envisaged that ribosome association with a leader sequence in the β-lactamase mRNA prevents the formation of a termination stem and loop structure and allows transcription to ensue. At low growth rates the cellular concentration of ribosomes is relatively low and transcription termination will occur. At faster growth rates there will be a higher concentration of ribosomes; these will associate with the leader sequence and allow transcription to proceed thus increasing the synthesis of β-lactamase.

Thus the repression of protease synthesis in microorganisms might be regulated in one of two ways. Repressor-operator control could be envisaged, with an apo-repressor that recognizes several amino acids. This seems unlikely, since the amino acids that cause repression are often structurally unrelated. Alternatively, attenuation may be employed which would require stretches of amino acid codons in a leader sequence that correspond to the amino acids that effect the repression. There are no experimental data to support either scheme at present.

Nitrogen regulation   Not only proteases but also many other enzymes that act on nitrogenous substrates, including urease, histidase, and periplasmic asparaginase, are repressed by ammonia or rapidly metabolized amino acids in a fashion analogous to carbon catabolite repression. This has been studied extensively in *A. nidulans* in which it is believed that the *areA* (ammonium regulation) locus encodes a protein that acts as a positive control element and activates those operons subject to ammonium repression. Its role seems to be to monitor the nitrogen status of the cell, and it is presumably inactivated by a plentiful, intracellular supply of nitrogen (probably glutamine). Should nitrogen become scarce, protease genes amongst others are activated by the *areA* product, probably at the transcriptional level.

Nitrogen regulation also occurs in prokaryotes, although the details are only understood in the enteric bacteria. Again glutamine appears to play a central role and, until recently it was thought that, when nitrogen was scarce, glutamine synthetase behaved as an activator of nitrogen regulated genes. When nitrogen was plentiful, the enzyme was chemically modified and lost its regulatory function. It is known that this attractive and simple scheme is largely incorrect and several other gene products which are encoded in the same operon as the glutamine synthetase structural gene, are involved in the activation and repression of nitrogen regulated operons (Magasanik, 1982). There is some doubt as to whether an analogous nitrogen regulation system occurs in Gram positive bacteria such as *Bacillus* but, if it does, the molecular mechanisms will probably be quite different.

Carbon catabolite repression   Despite the widespread occurrence of carbon catabolite repression, with the exception of enteric bacteria, the molecular details are poorly understood. In these Gram negative bacteria, cAMP together with its receptor protein (termed CRP or CAP) play a central role in catabolite repression. The CRP is a dimer of two identical subunits each capable of binding one molecule of cAMP, and it has two distinct domains. The N-terminal portion binds to cAMP and the COOH-terminal end binds to DNA. In the presence of cAMP, the protein undergoes a conformational change to an active configuration in which it preferentially binds to specific promoters enabling RNA polymerase to bind and initiate transcription at a second site 30 to 50 nucleotides upstream from the cAMP/CRP site. The specific promoters number about 30 in *E. coli*, and are responsible for genes involved in the transport and utilization of carbon fuel sources. Without this activation, even in the presence of inducers, transcription of catabolite sensitive operons is reduced to a very low level. Thus mutants of *E. coli* that lack CRP or adenyl cyclase (the enzyme that synthesizes cAMP from ATP) are unable to grow on most carbon sources except glucose or gluconate and were originally isolated through this pleiotropically negative phenotype.

During growth of *E. coli* on a rapidly metabolized carbon source, the intracellular concentration of cAMP is low and consequently the genes for various carbon catabolic enzymes are not expressed. The cAMP-CRP binding sites vary in sequence between different operons as do the RNA polymerase binding sites. There is evidence that these promoters form a hierarchy with increasing requirements for cAMP-CRP. Thus as the intracellular concentration of cAMP increases in response to carbon starvation, operons become activated in readiness for the relevant inducer molecule. Similarly, when glucose is added to bacteria growing slowly on lactose, there is a rapid decrease in the intracellular concentration of cAMP and catabolite sensitive operons are no longer transcribed resulting in the observed transient repression.

Despite intensive searches, cAMP and the enzymes responsible for its synthesis and degradation are undetectable in several *Bacillus* and *Lactobacillus* species, although they have been found in other Gram positive bacteria such as *Streptomyces, Nocardia* and *Arthrobacter*. Since *Bacillus* and *Lactobacillus* strains exhibit catabolite repression, the mechanism must be different from that in *E. coli* but at present no one has unravelled the details or intermediates involved.

The situation in eukaryotic microorganisms is no better. Most fungi exhibit glucose repression of a variety of intracellular and extracellular enzymes. *Saccharomyces* species have attracted considerable attention in this respect since they are able to utilize a variety of carbon sources for growth and they display catabolite repression. There has been a tendency to assume that cAMP is the mediator involved in these organisms, but the evidence is not convincing. Although *S. cerevisiae* contains twofold higher steady state levels of cAMP when grown on poor carbon sources compared with glucose, the kinetics of the increase in cAMP during the transition from glucose to another carbon source do not always correlate well with the appearance of catabolite repressible enzymes. Moreover, mutants of *S. cerevisiae* were obtained recently that would use cAMP to supplement a requirement for adenine. In these cAMP permeable strains, the presence of exogenous cAMP did not prevent glucose repression of galactokinase synthesis. Thus although yeast cells possess an entire cAMP-based regulatory system it seems to be involved in the regulation of cell division and differentiation rather than glucose repression. It is thought that this also applies to filamentous fungi.

Modulation of RNA polymerase activity Gross changes in transcriptional specificity can be achieved by modifying promoter recognition by RNA polymerase. Promoter recognition is directed by the σ factor and there is now firm evidence that changes of σ factor are involved in the process of sporulation in *B. subtilis*. At the onset of sporulation, there is a marked decrease in the activity of the 55 000 dalton σ factor and core RNA polymerase acquires various new polypeptides. It is known that at least one of these, a σ factor of molecular mass 37 000 preferentially transcribes sporulation-specific genes that are expressed early in sporulation. This is the first instance within a single bacterium of distinct σ factors (not associated with phage) that differ in promoter recognition specificities, and it presumably plays a key role in the developmental cycle of *B. subtilis*. Extracellular enzymes are often synthesized at a low rate during exponential growth and maximally during stationary phase. It has been suggested that this may reflect changes in RNA polymerase composition in that the vegetative σ factor initiates poorly at exoenzyme promoters, but the stationary phase σ factor has a strong affinity for these promoters. Thus, as the organism enters stationary phase, exoenzyme synthesis becomes maximal. Attractive as this hypothesis is, there is currently insufficient evidence either for its support or rejection.

RNA polymerase selectivity can also be modified by a variety of small molecules including various adenine and guanine nucleotides. For example *in vitro* experiments have shown that ATP inhibits transcription of *lac* operon DNA by RNA polymerase from *E. coli,* an effect that can be reversed by 5'-ADP or 5'-AMP. At the end of exponential growth, there are considerable changes in the intracellular concentrations of these small molecules and it may be that these are responsible for the observed patterns of extracellular enzyme synthesis. Whatever the mechanisms, however, from the previous accounts of nitrogen regulation and catabolite repression it will be apparent that the regulation of metabolism in different bacteria varies considerably and it will therefore be unlikely that a common mechanism will be responsible for the derepression of extracellular enzyme synthesis which occurs in many microorganisms.

Regulation of translation

Control of translation encompasses stimulation and repression of ribosome function and modulation of mRNA stability. Considerable evidence has accumulated to indicate that differential mRNA stability offers a measure of translational control over the synthesis of extracellular proteins.

One of the original findings in this context involved late exponential phase cells of *B. amyloliquefaciens* that had been washed and resuspended in fresh medium. Such cells continued to secrete amylase, protease and RN'ase for up to 90 min in the presence of sufficient rifampicin or actinomycin D to inhibit RNA synthesis. This secretion was readily inhibited by chloramphenicol or other inhibitors of translation, which suggested that the rifampicin resistant synthesis represented translation of a preformed message. Similar findings were subsequently reported for extracellular protein synthesis in other bacteria, both Gram positive and negative, and two explanations were offered. It could be that a large pool of mRNA accumulates during exponential growth and is translated during stationary phase when the rate of extracellular enzyme synthesis is maximal. This reservoir of mRNA would be sufficient to sustain extracellular enzyme synthesis in the absence of transcription for the observed period. Alternatively, the mRNA for these proteins may be abnormally stable, thus allowing continued translation in the absence of transcription. The kinetics of mRNA decay following the inhibition of transcription can be monitored as enzyme synthesis following treatment with rifampicin or some other inhibitor of transcription. Such experiments provide the 'functional' half-life of an mRNA species and for protease mRNA in *B. amyloliquefaciens* the half-life has been calculated to be about 9 min. Similarly, the half life for extracellular protease mRNA from *Bacillus megaterium* has been estimated at 6 to 7 min and for amylase mRNA from *B. licheniformis* around 8 min. This compares with an average mRNA half-life for cytoplasmic proteins of between 2.5 and 4 min. Thus it would seem that some extracellular enzyme mRNAs are relatively stable.

A similar study involving the pulse labelling of proteins synthesized after the inhibition of transcription provided functional half-lives between 40 sec and 20 min for mRNAs in *E. coli*. The more stable messages were associated with exported (cell envelope) proteins. Indeed, the high stability of the lipoprotein mRNA in *E. coli* has permitted its purification and subsequent sequence analysis. It consists of 322 nucleotides and has several unique features. In particular, it can form nine, stable stem and loop structures by base pairing and it is thought that these may be responsible for the unusual stability of the molecule (Inouye, 1982). Alternatively, it is tempting to speculate that the stability of these mRNAs results from the presence of bound ribosomes that are temporarily halted while the polysome migrates to the membrane where cotranslational secretion will ensue (Chapter 2). Attractive as this supposition is, recent studies revealed no difference in the stabilities of translated versus untranslated mRNAs for β-lactamase nor OmpA (outer membrane) protein in *E. coli*. Thus the molecular basis for the greatly (5 to 10-fold) increased stability of some exported protein mRNAs remains to be elucidated as does the precise mechanism of mRNA degradation.

Functions of exoenzymes

Most extracellular enzymes have no cellular substrate and it would appear that they have evolved as 'scavenger' enzymes which degrade polymeric material in the environment to provide the organism with assimilable nutrients. However, some extracellular enzymes have cellular substrates, in particular proteases, nucleic acid hydrolysing enzymes and cell wall lytic enzymes. These have been implicated in several cellular processes, in particular differentiation and cell wall turnover and growth. This impinges on the regulation of their synthesis which will be associated with the overall processes involved, rather than merely induction and repression by molecules in the environment.

Differentiation In *Bacillus*, maximal synthesis of exoenzymes often occurs prior to sporulation in the late exponential growth phase. In early studies it was suggested that these two events may be connected and some extracellular enzymes may be involved in sporulation. However, since those environmental conditions that control extracellular enzyme synthesis also control sporulation, it is difficult to decide if exoenzyme synthesis is due to altered environmental conditions or the expression of a sporulation-specific gene. In many instances mutants have clarified the situation. Clearly, if a mutant is deficient in a particular enzyme and yet sporulates normally this enzyme can play no part in sporulation. In this way the amylase and neutral protease of *B. subtilis* were shown to be independent of sporulation, since Amy⁻ and Npr⁻ mutants sporulated normally. The situation was somewhat more complicated with the serine protease of *B. subtilis*, since all mutants defective in the synthesis of this enzyme showed defects in sporulation. This does not necessarily imply a function for serine protease in sporulation however, but merely indicates that the two events are coordinately regulated. A direct relationship can only be certified if the sporulation defective mutation can be assigned to the structural gene of the enzyme concerned, and this has now been demonstrated for those extracellular enzymes in *Bacillus* likely to be involved in sporulation.

It seems likely that certain steps in sporulation are necessary for the triggering of synthesis of some extracellular enzymes. Amylase appears to be an exception and is synthesized normally when sporulation is blocked, but the synthesis of serine protease is invariably affected when sporulation is inhibited by mutation or by thymine starvation. The neutral protease and RN'ase are generally, but not always, affected by defects in sporulation. It may be concluded therefore, that although extracellular enzymes are not directly involved in sporulation, the synthesis of many of these enzymes is dependent on the sequence of events leading to the formation of the endospore, and some sporulation specific genes must be expressed before the enzymes can be synthesized normally.

A similar situation prevails in eukaryotic microorganisms. Although variation in protease activity often accompanies differentiation in yeasts and moulds, these are generally intracellular enzymes. Indeed, mutants of *A. nidulans* lacking extracellular protease develop normally. The β-glucanases of *S. cerevisiae* have received considerable attention in this context. The levels of various glucandegradative enzymes alter significantly upon sporulation but, like the amylase of *B. subtilis*, this is mainly due to changes in the cultural conditions. Nevertheless, inhibitors of sporulation prevent the synthesis of a particular extracellular 1,3-β glucanase, the appearance of which parallels the formation of asci. It may be that this enzyme is involved in the processing of glucan which appears at this time or, like the serine protease of *B. subtilis*, its synthesis may simply be triggered by differentiation. This point could be clarified by genetic studies as described above.

Growth and cell division Extracellular enzymes that hydrolyse molecules within the cell wall (autolysins) have an obvious role in growth and cell division. Autolysin deficient cells generally grow and partition themselves in the normal fashion, but cannot separate progeny effectively. These mutants often assume bizarre morphologies, but the fact that they can grow suggests that the missing activities are associated with division-related wall cleavage rather than growth. However, all of the autolysin-deficient mutants isolated to date retain low levels of activity. This indicates that autolysin activity is necessary for local wall cleavage preceding insertion of new polymer during growth, and mutations resulting in a complete deficiency of autolysin would be lethal. At present it is therefore impossible to come to any firm conclusions about the role of autolysins in wall growth and cell division.

Summary

The synthesis of extracellular enzymes is regulated by the environment in the same way as cytoplasmic enzymes. If inducible, they are generally induced by a product from the substrate rather than the substrate itself since the latter is usually too large to enter the cell. Accumulation of the products generally leads to repression of the enzyme. Present evidence suggests that the control is exerted primarily at transcription and it seems probable that inducible extracellular enzymes are governed by repressor or activator proteins. Thus the operon model for gene regulation is applicable to these systems and has been confirmed in the case of penicillinase synthesis in *B. licheniformis*. Extracellular enzyme synthesis is repressed by rapidly metabolized carbon sources (catabolite repression). Although cAMP is the major mediator of catabolite repression in enteric bacteria it should be noted that this molecule is absent from those *Bacillus* strains examined to date and the mechanism(s) of catabolite repression in many bacteria remains unknown. Similarly, an ample supply of nitrogen as amino acids or ammonium represses the synthesis of proteases and related enzymes.

A particular feature of many extracellular enzymes is the restriction of synthesis to the late exponential and early stationary growth phases. This was originally thought to imply that these enzymes had a function in differentiation (for example sporulation); however, mutants that lack exoenzymes, but that sporulate normally, have disproved this. Nevertheless, it seems likely that some event involved in differentiation may be necessary to trigger the synthesis of some extracellular enzymes; others such as amylases are completely divorced from differentiation. The derepression of these enzymes after exponential growth probably results from modification of RNA polymerase specificity through modulation of some small effector molecule or through changes in RNA polymerase composition. A second peculiarity of many of these proteins is the extended stability of their mRNAs. This may be associated with the need to transport the mRNA or polysome from its site of synthesis to translation sites on the inner surface of the cytoplasmic membrane.

References

DARNELL, Jr., J. E. (1982). Variety in the level of gene control in eukaryotic cells. *Nature* 297: 365–71.

BOTSFORD, J. L. (1981). Cyclic nucleotides in prokaryotes. *Microbiological Reviews* 45: 620–42.

von GABAIN. A., BELASCO, J. G., SCHOTTEL, J. L., CHANG, A. C. Y. and COHEN, S. C. (1983). Decay of mRNA in *Escherichia coli*: investigation of the fate of specific segments of transcripts. *Proceedings of the National Academy of Sciences USA* 80: 653–7.

IMANAKA, T., TANAKA, T., TSUNEKAWA, H. and AIBA, S. (1981). Cloning of the genes for pencillinase, *penP* and *penI*, of *Bacillus licheniformis* in some vector plasmids and their expression in *Escherichia coli*, *Bacillus subtilis* and *Bacillus licheniformis*. *Journal of Bacteriology* 147: 776–86.

INOUYE, M. (1982). Lipoproteins from the bacterial outer membranes: their gene structures and assembly mechanism. In: *Membranes and Transport* Vol. 1, pp. 289–98. Edited by A.N. Martonosi. Plenum Press, New York.

LOSICK, R. (1982). Sporulation genes and their regulation. In: *Molecular Biology of the Bacilli* Vol. 1, *Bacillus subtilis*, pp. 179–201. Edited by D.A. Dubnau. Academic Press, New York and London.

MAGASANIK, B. (1982). Genetic control of nitrogen assimilation in bacteria. *Annual Review of Genetics* 16: 135–68.

PALL, M. L. (1981). Adenosine 3'-5'-phosphate in fungi. *Microbiological Reviews* 45: 462–80.

PRIEST, F. G. (1977). Extracellular enzyme synthesis in the genus *Bacillus*. *Bacteriological Reviews* 41: 711–53.

SAVAGEAU, M. A. (1979). Autogenous and classical control of gene expression: a general theory and experimental evidence. In: *Biological Regulation and Development*. Vol. 1. *Gene Expression*, pp. 57–108. Edited by R. F. Goldberger. Plenum Press, New York.

YANOFSKY, C. and KOLTER, R. (1982). Attenuation in amino acid biosynthetic operons. *Annual Review of Genetics* 16: 113–34.

4 Commercial enzymes

The sources of the most important industrial enzymes are shown in Table 2. Relatively few bacterial genera are involved, the aerobic, endospore-forming rods of the genus *Bacillus* being predominant. Only a single Gram negative bacterium, *Klebsiella pneumoniae* (*'Klebsiella aerogenes'*) is used, which reflects the paucity of true extracellular enzymes in these bacteria and the difficulty and expense of recovering periplasmic enzymes on a commercial scale. With regard to the eukaryotes, the filamentous fungi *Aspergillus*, *Penicillium* and *Rhizopus* together with some yeasts are commonly used.

Starch-hydrolysing enzymes

Conversion of starch into sugars, syrups and dextrins forms the basis of the starch-processing industry. These hydrolysates are used in a range of manufactured foods and beverages and as carbon sources in fermentations. Starch hydrolysates were originally prepared by acid hydrolysis but enzymic processes offer several advantages and now dominate the industry. The benefits are: (1) fewer byproducts are formed and off-flavours are therefore avoided, (2) enzymes are specific, and consequently homogeneous end-products with desired physical properties can be obtained e.g. osmotic pressure, sweetness or resistance to crystallization and (3) higher product yields are obtained with enzymes since caramelization is avoided. Before considering the enzymes and processes it will be useful to outline the structure of the substrate, starch.

Starch   Starch occurs in water-insoluble granules, the shape and size of which are often characteristic of the plant. It constitutes the major reserve carbohydrate in all higher plants and is produced commercially from seeds, tubers and roots. The major source in the Western hemisphere is corn (maize); in developing countries cassava tubers and the stem of the sago plant are important sources.
Starch comprises two high molecular weight components (Figure 9), amylose and amylopectin. These differ significantly in several physical properties, in particular solubility in water and molecular size. Amylose is a linear chain of around 10^3 glucose residues linked through 1,4-α-bonds and is poorly soluble in water. Amylopectin is a highly branched polymer of 10^4 to 10^5 glucose residues and comprises 1,4-α-linked glucose chains (20 to 25 residues long) linked at 1,6-α branch points. The molecule therefore contains 4 to 5% of 1,6-α-linkages, is soluble in water and accounts for 75 to 85% of starch depending on the source of the material.

α-**Amylase** α-Amylase is an endo-attacking enzyme that hydrolyses the 1,4-α-linkages in amylose and amylopectin. This random attack results in a rapid reduction in iodine-staining capacity and viscosity of starch solutions. Hydrolysis of amylose yields low-molecular weight oligosaccharides ranging from the

Amylopectin

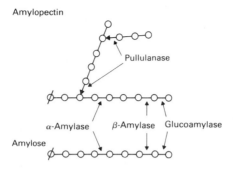

Amylose

Fig. 9 Schematic structures of portions of amylopectin and amylose showing bonds hydrolysed by various enzymes. ○, glucose residue; ∅, glucose with an exposed reducing group. →1,6-α-linkage; —, 1,4-α-linkage.

disaccharide maltose, to maltohexaose depending on the origin and nature of the amylase. Hydrolysis of amylopectin results in glucose, maltose and a variety of branched α-limit dextrins that contain the 1,6-α-linkages from the original polymer.

Bacillus strains are used for most of the commercial production of α-amylase but the species involved have been the subject of considerable confusion. In 1945, Fukumoto isolated a bacterium from soil that secreted large quantities of α-amylase and other enzymes. He consequently named this organism *Bacillus amyloliquefaciens* but it closely resembles *Bacillus subtilis* (which also secretes α-amylase) and a lack of phenotypic attributes with which to distinguish these bacteria led to *B. amyloliquefaciens* being considered synonymous with *B. subtilis*. This situation promoted considerable confusion amongst enzymologists and physiologists since the two organisms are genetically and physiologically distinct and secrete α-amylases that are molecularly and serologically different. More recently, *B. amyloliquefaciens* has been afforded species status in recognition of its lack of affinity with *B. subtilis*.

B. amyloliquefaciens is used for much of the commercial production of bacterial α-amylase. The enzyme has a molecular weight of about 50,000 daltons and is stable in the pH range 5.5 to 8.0 with optimum activity at pH 5.9. Like other α-amylases it is a calcium metalloenzyme, the metal being required both for activity and for increased stability. If Ca^{2+} is added to the reaction mixture, the enzyme is quite heat stable and can be used for starch hydrolysis at temperatures approaching 90°C. However, the starch industry would like to use higher hydrolysis temperatures. This is because starch granules are resistant to α-amylase attack and must be ruptured by heat before hydrolysis can occur. 'Cooking' temperatures above 100°C are normally required to ensure complete breakage of maize granules and gelatinization of the starch. An enzyme that could operate at this higher temperature would enable faster, more efficient and cheaper starch liquefaction since the material would not have to be cooled after gelatinization for enzymic hydrolysis.

The α-amylase from *B. licheniformis* was introduced in 1972. It has remarkable heat stability and can be used for starch hydrolysis at gelatinization temperature (105 to 110°C). The *B. licheniformis* enzyme has a wider pH range for both stability and activity than the *B. amyloliquefaciens* enzyme. It is larger (60,000

33

daltons) and, despite being more temperature stable, is less dependent on Ca^{2+}. The enzyme has only one disadvantage: it is very difficult to terminate the reaction by heat treatment! Despite this shortcoming, the *B. licheniformis* enzyme has captured a large sector of the world market for α-amylase over the past decade.

α-Amylase is also prepared commercially from *A. oryzae*. This enzyme differs from the *Bacillus* α-amylases in being a glycoprotein. Moreover, it has a lower temperature stability (50 to 55°C) and a lower pH range for activity and stability than the bacterial enzymes. *A. oryzae* amylase is more saccharifying than its bacterial counterparts in that it hydrolyses starch to lower molecular weight oligosaccharides. It is possible to obtain 50% of maltose from starch with this enzyme and it is used for the commercial production of high-maltose syrups and in the baking industry.

Glucoamylase Glucoamylase, also known as amyloglucosidase, is an exo-acting enzyme that yields β-D-glucose by hydrolysing 1,4-α-linkages consecutively from the non-reducing chain ends of amylose and amylopectin. It also hydrolyses 1,6-α-linkages, albeit at a slower rate. The rate of hydrolysis increases with the size of the substrate to a maximum when the substrate has five or more monosaccharide units. The enzyme is therefore misnamed as amyloglucosidase since glucosidases do not invert product configuration as this enzyme does by producing β-glucose from an α-linked substrate, and hydrolyse oligosaccharides at lower rates than disaccharides.

Glucoamylases are rare in bacteria but have been found in several genera of fungi. The enzymes from *Aspergillus*, *Rhizopus* and *Endomyces* have been produced commercially but the *Aspergillus* enzyme dominates the market since it is the most thermostable. The optimum temperature is 60°C rather than 55°C, an important consideration since it is difficult to prevent microbial growth in the reaction mixture at temperatures below 60°C. Glucoamylase from *A. niger* has been characterized and at least two enzyme components have been purified from culture broth. Both are glycoproteins containing 13 to 18% carbohydrate and have molecular weights around 100,000 daltons. Their optimum pH for activity is about 4.5.

An interesting feature of many glucoamylases is their ability to attack and hydrolyse raw starch granules. With the high cost of energy this has assumed increased significance since it offers the possibility of hydrolysing·starch without the energy intensive, high temperature gelatinization stage. Such processes have been introduced on a small industrial scale.

β-Amylase β-Amylase occurs widely in higher plants and the enzyme from malted barley is used in the brewing and distilling industries for the hydrolysis of starch into fermentable sugars. The enzyme degrades amylose and amylopectin in an exo-fashion from the non-reducing chain ends by hydrolysing alternate 1,4-α-linkages to yield maltose in the β-anomeric form. Since the enzyme can neither attack nor hydrolyse 1,6-α-linkages, a high molecular weight β-limit dextrin is produced from amylopectin. β-Amylase has only recently been characterized as an extracellular enzyme from bacteria; producing organisms include *Bacillus polymyxa*, *Bacillus megaterium*, *Bacillus cereus*, *Pseudomonas* and *Streptomyces* strains. Microbial β-amylases, like the plant enzymes, are thiol enzymes that have no metal ion requirements and are inactivated by *p*-chloromercuribenzoate and by oxidation. Although there is an extensive market for β-amylases in brewing and distilling and for the manufacture of high-maltose syrups, the low temperature stability of the currently-known enzymes (45 to 55°C) precludes their use.

Debranching enzymes Debranching enzymes hydrolyse 1,6-α-linkages in amylopectin and/or glycogen. Only one such enzyme is produced commercially, pullulanase which gains its name by virtue of its attack on a fungal polysaccharide, pullulan. Pullulan is liberated by *Aureobasidium pullulans* and essentially comprises maltotriose units linked by 1,6-α-bonds. Pullulanase hydrolyses these 1,6-α-bonds and the 1,6-α-branch points in amylopectin. Pullulanase is widely distributed in both Gram positive and negative bacteria but only the enzyme from *K. pneumoniae* is synthesized in sufficient quantity to render commercial production economic. This is therefore one of the very few commercial enzymes prepared from a Gram negative bacterium. The enzyme is located in the periplasm and the outer membrane of the bacterium but is also released into the culture fluid, particularly by high-yielding strains. The enzyme has activity optima of 50°C and pH 6.5; the low temperature optimum limits its industrial usage, which is to assist in the total degradation of starch to glucose.

Glucose isomerase Although it is not a starch-hydrolysing enzyme, glucose isomerase (D-xylose betol-isomerase; xylose isomerase) is included here because it is used to convert glucose, derived from starch, into fructose. This conversion is commercially valuable because of the increased sweetness, depending on application and conditions, of the product. Glucose is around 70% as sweet as sucrose whereas fructose is about 50% sweeter than sucrose. Thus high fructose corn sweeteners (HFCS) can be used to replace sucrose in beverages and foods. Production of HFCS in USA has developed from virtually zero in 1970 to 10% of the entire production of calorific sweeteners in 1980. The price advantage of HFCS in USA derives from an abundance of cheap corn, efficient processing techniques and good markets for the byproducts (corn oil and protein). The second generation of HFCS containing 55 and 90% fructose, compared to the original 42%, predict a bright future for this technology, particularly in the beverage industry. For example Coca Cola Co. recently announced that sucrose has been largely replaced by HFCS in its name brand.

Fructose syrup is also manufactured in Japan and Europe but in EEC countries the strong political influence of the agricultural industry and the sugar beet farmers has severely hindered its development and use.

The successful development of HFCS stemmed from the discovery of glucose isomerizing enzymes in bacteria and the development of suitable technology to utilize them efficiently. Of the four types of microbial glucose isomerizing enzymes, only D-xylose isomerase has commercial usage since it is relatively heat stable (capable of reacting in the range 45 to 65°C) and does not require any cofactors. The enzyme is widely distributed and is produced by most microorganisms capable of growth on xylose. The sources of the commercial enzymes are shown in Table 2. The enzymes from *Lactobacillus brevis*, *Streptomyces albus* and *Bacillus coagulans* have been studied in some detail. They are of a similar size (191,000; 165,000 and 167,000 daltons respectively) and comprise four identical subunits. The *L. brevis* enzyme has a low temperature optimum (45°C) whereas most xylose isomerases have optima above 65°C and as high as 100°C for some enzymes from thermophiles. The pH optima are generally above neutrality. Xylose isomerase catalyses the reversible reaction: D-xylose $\rightleftharpoons$ D-xylulose.

Extracellular Enzymes

Table 2 Sources of important commercial enzymes

Enzyme	Producer-organisms
α-Amylase	*Aspergillus oryzae*
	Bacillus amyloliquefaciens
	Bacillus licheniformis
Cellulase	*Aspergillus* sp
	Trichoderma reesei
Dextranase	*Penicillium* sp
β-Glucanase	*Aspergillus niger*
	Bacillus amyloliquefaciens
Glucoamylase (amyloglucosidase)	*Aspergillus niger*
	Rhizopus sp
Glucose isomerase	*Actinoplanes missouriensis*
	Arthrobacter sp
	Bacillus coagulans
	Streptomyces species
Hemicellulase	*Aspergillus niger*
Invertase	*Aspergillus* sp
	Saccharomyces sp
Lactase	*Aspergillus niger*
	Kluyveromyces marxianus
	(*"Saccharomyces fragilis"*)
	Kluyveromyces lactis
Lipase	*Aspergillus* sp
	Candida cylindracea
	Mucor miehei
	Rhizopus sp
Pectinase	*Aspergillus niger*
Protease	*Aspergillus niger*
	Aspergillus oryzae
	Bacillus amyloliquefaciens
	Bacillus licheniformis
	Bacillus stearothermophilus
	(*"B. thermoproteolyticus"*)
	Alkalophilic *Bacillus* sp
Pullulanase	*Klebsiella pneumoniae*
	(*"K. aerogenes"*)

During growth the xylulose is then phosphorylated and enters the pentose phosphate metabolic pathway after conversion to ribulose-5-phosphate. All known xylose isomerases are also thought to catalyse the conversion of D-glucose to D-fructose. The specificities to other sugars vary with the source of the enzymes. The reaction follows simple Michaelis–Menton kinetics and in general the Km is much lower and V_{max} higher for xylose than glucose. Hence these enzymes are thought of as xylose isomerases. Since the reaction is reversible, an equilibrium is reached which determines the practical upper limit of conversion.

At equilibrium the fructose concentration varies from 50% at 60°C to almost 55% at 85°. Current research is principally directed towards ways of improving the fructose concentration in the isomerized product through complexing the fructose or shifting the glucose-fructose equilibrium by varying reaction conditions.

Applications and usage of starch-hydrolysing enzymes During the past 20 years, the hydrolysis of starch to produce syrups of various compositions has become one of the most important of all industrial processes using enzymes. The hydrolysates are classified according to their content of reducing sugars represented as glucose and described as DE (dextrose equivalent), pure glucose being DE100 and starch near zero. Selected combinations of enzymes and conditions yield products with well-defined physical and chemical properties. The basic procedures are shown in Figure 10. After milling of the source material the starch is dispersed or gelatinized in aqueous solution and thinned or liquefied wih a thermostable bacterial α-amylase at temperatures ranging from 80 to 110°C. Acid hydrolysis may be substituted for this step but is generally less efficient. Liquefaction takes 2 to 4 hours and is normally terminated when the DE is 10 to 20.

Liquefaction is followed by saccharification, a procedure which can be varied depending on the desired product. Addition of the debranching enzyme pullulanase and β-amylase to the dextrins will yield a high-maltose syrup which currently has a limited market but has some particular advantages in the food industry in that it does not crystallize and is relatively non-hygroscopic. Since suitable microbial β-amylases remain scarce, high-maltose syrups are generally prepared with mixtures of fungal α-amylase and glucoamylase. This material will contain more glucose but the ratio of glucose to maltose can be controlled by carefully adjusting the proportions of the two enzymes and the reaction conditions.

The main commercial process is the production of glucose for isomerization to fructose. After liquefaction, the dextrins are flash-cooled to about 60°C and treated with glucoamylase for 24 to 90 hours depending on the amount of enzyme used. The saccharified starch should contain at least 94 to 96% glucose since remaining di- and oligosaccharides often have unacceptable taste and none are affected by the isomerase. After adjustment of pH and ionic strength the material is treated with an immobilized isomerase (see page 46). Fructose syrups usually comprise 42% fructose which requires an initial glucose level in the substrate of about 94% on a dry weight basis.

Another important application of the bacterial α-amylases includes the

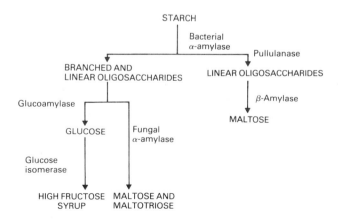

Fig. 10 Commercial processes of starch hydrolysis.

liquefaction of starchy raw materials for the production of alcohol both as a fuel and in the brewing industry. Fungal α-amylase is used in the baking industry for the degradation of starch in the dough. The maltose produced is metabolized by the yeast during leavening and the low temperature stability of the enzyme ensures that it is rapidly removed during baking. Glucoamylase is used extensively in the production of alcohol from starchy raw materials.

Proteases

Proteolytic enzymes are ubiquitous in microorganisms and the extracellular proteases are probably the most widespread of all microbial, secreted enzymes. These enzymes are simple to detect and are often synthesized in high yield. They have therefore been extensively studied and their molecular properties are understood in detail. Microbial proteases may be grouped into three classes; serine, metallo and acid proteases which have alkaline, neutral and acid pH optima, respectively. Serine and metalloproteases are produced commercially from *Bacillus* strains while the acid proteases are prepared from fungi, notably *Mucor* and *Endothea* strains.

Serine protease   Serine proteases characteristically possess a serine residue at the active site of the molecule. They are endopeptidases with a strong proteolytic activity coupled with low specificity. They are stable molecules with an alkaline pH optimum (9.0 to 11.0) and a molecular weight of 25,000 to 30,000 daltons. Serine proteases are inhibited by diisopropylfluorophosphate and phenylmethyl sulphonyl fluoride but not by chelating agents. Generally they resemble the animal enzyme trypsin.

The first microbial protease to be obtained in crystalline form (in 1952) was derived from a *Bacillis* strain discovered by Lang and Ottetson in the Carlsberg Laboratory in 1947. Unfortunately the taxonomy of the genus *Bacillus* was uncertain at that time and the organism was incorrectly identified as *B. subtilis*. The enzyme consequently became known as subtilisin Carlsberg. It is now clear that the organism was not *B. subtilis* but the closely-related *B. licheniformis*. Comparison of subtilisin Carlsberg with serine proteases from other strains of *B. licheniformis* shows the enzymes to be essentially identical. These proteases have no metal ion requirement for activity and, in common with most serine proteases, Ca^{2+} is not necessary for stability. This is a particularly relevant feature since serine proteases are used extensively in washing detergents that contain sequestering agents such as EDTA or tripolyphosphate. The enzyme is stable between pH 5 and 11 and at 50°C for 1 hour. It is rapidly inactivated above 70°C and by oxidizing agents such as hypochlorite and hydrogen peroxide.

The search for temperature and alkaline-stable proteases for inclusion in washing detergents involved the isolation of *Bacillus* strains from the environment that would grow to pH values above 10. Such strains are common and although some have been given species designations (e.g. *B. alcalophilus*) most seem to be alkalophilic variants of common species. The serine proteases from these bacteria have properties similar to subtilisin Carlsberg but are often more temperature stable and generally active over a higher pH range (6 to 12). In detergents they are often superior to subtilisin Carlsberg and their high pH stability has made possible a new process for animal hide dehairing.

Use of serine proteases in detergents Proteases have been used in washing detergents on a large scale for some 15 years and the current annual production of serine proteases in this context amounts to about 500 tonnes pure protein. Proteases are obviously useful for cleaning clothes soiled with blood, foods or some other proteinaceous matter but it is less generally realized that they can also improve overall washing efficiency.

Washing detergents comprise surface-acting agents, sequestering agents and alkali. In Europe, perborate is also commonly added and releases hydrogen peroxide, which bleaches the fabric, at temperatures above 50°C. The 'dirt' on the clothes comprises inorganic material often adhering to the fabric with protein derived from skin, food and plant materials. During laundering, the detergent disperses and dissolves much of the 'dirt' but proteins often coagulate on the fabric and are not dispersed. Serine protease hydrolyses this protein thus releasing residual 'dirt' and improving laundering efficiency. Enzyme washing powders should ideally be used for presoaking clothes at a relatively low temperature but this is not always possible. Nevertheless the enzyme functions effectively up to 60°C so long as perborate is not present. Although the enzyme is inhibited by cationic surfactants, it is relatively resistant to non-ionic and anionic surfactants. Since the enzyme has no metal ion requirement, sequestering agents in the washing powder are not harmful. Washing powders typically contain 0.015% active enzyme protein which provides an enzyme concentration in the suds of about 0.75 μg ml^{-1}. This may appear to be very low but serine protease has a high affinity for protein and operates effectively. Proteases from alkalophilic bacilli tend to be superior to subtilisin Carlsberg not only in their higher temperature and alkaline stability but also because they often have a still higher affinity for proteinaceous dirt. This is explained by their larger ionic charge; the pI for subtilisin Carlsberg is 9.4 and about 11.0 for highly alkaline proteases.

Enzyme washing detergents were quickly accepted by the market but there followed a setback when production workers developed allergies to enzyme dust. This has been remedied by introducing dust-free enzyme preparations. It is predicted that enzyme washing powders will be more widely used as the phosphate concentration is reduced to save resources.

Use of serine proteases in the leather industry Proteases have been used in two processes in the leather industry; removing hair from the hides and bating, a technique which softens leather. Dehairing hides has traditionally been accomplished by treatment with lime and sulphide. This procedure is discouraged because of the possibility of the production of toxic hydrogen sulphide, as well as problems with the disposal of the waste water which contains sulphide and hair, and is highly alkaline. The enzymic processes involving lime to swell and soften the hide and alkali-stable, serine protease from an alkalophilic *Bacillus* strain to loosen the hair have proved efficient, but unfortunately the relatively high cost of the enzyme has hindered their commercial implementation. With hair removed, the hides are softened in a process that remains chemically obscure. Bating has traditionally involved pancreatic enzymes but protease preparations from *A. oryzae*, *B. amyloliquefaciens* and *B. licheniformis* are replacing the animal enzymes.

Metalloproteases The metalloproteases typically contain an essential metal atom, usually zinc, and show optimal activity at neutral pH. Moreover, Ca^{2+} is essential for stability and the molecules are therefore inactivated and destabilized by chelating agents. They are endo-acting enzymes that preferentially cleave peptides with hydrophobic side-chains.

Metalloproteases are widespread and are synthesized by several bacilli including *B. amyloliquefaciens, B. cereus, B. megaterium, B. polymyxa* and *B. subtilis.* An exceptionally thermostable version (thermolysin) is secreted by '*B. thermoproteolyticus*' (*B. stearothermophilus*). This enzyme loses 50% of its activity in 15 min at 84°C compared with a similar loss at 59°C for the enzyme from *B. amyloliquefaciens.* The reason for this temperature stability is largely unknown since the amino acid sequences of the two molecules share extensive homology. It has been suggested that the higher proportion of hydrophobic residues and two additional Ca^{2+} ions in thermolysin might be responsible. Commercially, metalloproteases are prepared from *B. amyloliquefaciens* which is grown under conditions that repress serine protease and α-amylase synthesis, and from '*B. thermoproteolyticus*'.

Uses of metalloproteases The metal protease of *B. amyloliquefaciens* is used for leather bating as mentioned above and also finds application in the brewing and alcohol industries. It can be used to prevent chill haze, a proteinaceous precipitate that forms during cold storage of beer. The plant enzymes papain and bromelain are also used for this purpose.

For both potable and fuel alcohol production it is common to replace malted barley wth unmalted cereal and other starchy adjuncts. It is then necessary to supplement the mash with both α-amylase and protease to supply fermentable sugars and amino acids. The metal protease of *B. amyloliquefaciens* is ideal for this purpose since any contaminating α-amylase or β-glucanase secreted by *B. amyloliquefaciens* and present in the product is beneficial: the pH activity range is suitable and the enzyme is unaffected by barley serine protease inhibitor.

Acid proteases Proteases with low pH optima are rare in bacteria and predominate in fungi. These molecules invariably contain an aspartic acid residue at the active site and are unaffected by chelating agents, thiol-group reagents or by serine protease inhibitors. They share considerable affinity with the animal digestive enzymes pepsin and rennin. These enzymes are therefore used as replacements for animal proteases and major applications involve cheese and soy sauce manufacture. Several acid proteases are marketed for industrial usage, the most successful one being derived from *Mucor, Endothea* and *Aspergillus* strains.

The *Mucor* proteases are used extensively for cheese manufacture, an application that was suggested in the 1920s, but attempts were unsuccessful at that time because the enzymes available were unsuitable. Studies in the 1960s revealed two sources of appropriate enzymes, both thermophilic *Mucor* strains, *M. pusilus* and *M. miehei* capable of growth at 55° and 60°C, respectively. The *M. pusilus* enzyme is a typical acid protease comprising a single polypeptide chain (30,000 daltons) with no carbohydrate. The enzyme is specific for peptide bonds with aromatic side chains, is stable between pH 3 and 6 and loses 90% of its activity after 15 min at 65°C. During growth, *M. pusilus* secretes other undesirable enzymes including lipase and a non-specific protease and these must be removed before the enzyme can be used for cheese manufacture. The lipase is generally destroyed by acid treatment and the non-specific protease by adsorption on silicon dioxide or something similar.

The protease from *M. miehei* is similar to the *M. pusilus* enzyme but is slightly larger (38,000 daltons) and contains about 6% carbohydrate. Its milk coagulating ability is also similar to the *M. pusilus* protease but differences in dependency on temperature and Ca^{2+} concentration enable it to be used for alternative cheese manufacturing processes.

Endothea parasitica protease was marketed as a microbial rennet some 14 years ago. The enzyme is stable at pH 4 to 4.5 and inactivated in less than 5 min at 60°C. It has found limited acceptability as a rennet substitute because it has high proteolytic activity relative to milk-coagulating activity. Only in the production of Emmenthal cheese has it proved superior to *Mucor* rennet substitutes.

Aspergillus oryzae and the closely related *A. sojae* are important sources of commercial enzymes. *A. oryzae* is used to prepare Takadiastase, a traditional multienzyme preparation that contains acid, neutral and alkaline proteases and α-amylase with varying amounts of glucoamylase, cellulase and pectinase. It is used in the baking industry but its principal use is as a digestive aid. Similar enzymes from *A. sojae* are used to hydrolyse soy bean protein in the manufacture of soy sauce.

Uses of acid proteases Microbial enzymes have long been considered as a replacement for calf rennet. In the manufacture of cheese, milk proteins are coagulated to form a solid curd. This coagulation is usually achieved by lowering the pH of the milk to the isoelectric point of casein by the use of a starter culture of lactic acid bacteria, by using milk-clotting enzymes or by a combination of both of these procedures. Milk-clotting enzymes should preferably possess a high coagulating activity relative to proteolytic activity. The coagulating process involves the formation of a Ca^{2+}-linked casein complex and obviously if there is a high proteolytic activity present, the curd is degraded. Rennet, an extract from the stomach of young cows, contains chymosin (rennin) which is ideal for this process but as the animal ages the rennin is replaced by pepsin. Since pepsin is not satisfactory for cheesemaking, calf rennet must be used and this is in short supply because young cows are not slaughtered so frequently today.

Their unlimited supply, cheap manufacture and consistent quality strongly recommend microbial rennets. The ratio of milk-clotting to proteolytic activity for calf rennet is 500 to 1000 to 1, for *Mucor* rennet about 300 to 1 and for *E. parasitica* rennet 80 to 1 so the *Mucor* enzymes have a clear advantage. These enzymes have been used successfully for the manufacture of most types of cheese after suitable modification to the process and they currently command about 10% of the rennet market. Indeed, the potential market for microbial rennets is sufficiently important to have prompted several companies into cloning the relevant genes into alternative hosts thereby providing higher yields and removing the necessity to purify the enzyme.

Fungal proteases also find application in the baking industry. The gluten content of flours used for bread production can fluctuate widely, necessitating variation in dough manufacture to provide a consistent bread. Fungal proteases are used to degrade the gluten to a constant level, thus allowing a standard baking operation and generally shorter mixing and resting periods. Protease from *A. oryzae* is commonly used since it is active at the pH of dough and is rapidly degraded by the heat of baking.

Other enzymes

The proteases, starch-hydrolysing enzymes and glucose isomerase together command about 90% of the industrial enzyme market. Nevertheless there are several other enzymes that currently have important uses and are prepared in bulk for the industrial consumer or have considerable potential application.

Cellulase   Cellulose is the principal component of plant cell walls and one of the most abundant biological compounds. It is a major waste product both in nature and from man's activities, and the possibility of hydrolysing this material to glucose is immensely attractive. However, economic processes for this conversion have yet to be devised despite considerable research effort.

Cellulose is a linear polymer of 1,4-β-linked glucose residues, the number of which varies considerably but averages some 3000. Individual cellulose molecules are linked together by hydrogen bonds to give larger, crystalline structures although the number of molecules in such units and their organization are not definitely known. Because of this complex structure, crystalline cellulose is not amenable to attack by single enzymes and cellulolytic microorganisms secrete a variety of cellulases to deal wth the composite molecule.

Enzyme systems involved in cellulose degradation have been thoroughly studied in two fungi, the white-rot fungus *Sporotrichum pulverulentum* and the mould *Trichoderma reesei* (previously *T. viride* QM6a, now renamed *T. reesei*). White-rot fungi have strong wood-degrading capability and can degrade lignin in addition to cellulose. The fungus normally grows in the cell lumen and the hyphae penetrate from one cell to another through naturally-occurring openings or by boring holes in the cell walls. Scanning electron micrographs of fungal growth on wood have revealed the disappearance of lignin and cellulose remote from the hyphal wall, presumably due to diffusable extracellular enzymes.

Enzymic hydrolysis of cellulose by *S. pulverulentum* and *T. reesei* is similar. *S. pulverulentum* secretes three classes of cellulolytic enzymes: (1) five different 1,4-β-endoglucanases which randomly attack the 1,4-β-linkages in the polymer, (2) a 1,4-β-exoglucanase which removes cellobiose or glucose residues from the non-reducing chain end of the molecule and (3) two 1,4-β-glucosidases which hydrolyse cellobiose and water soluble cellodextrins to glucose. The enzyme system from *T. reesei* also comprises endo- and exoglucanases and β-glucosidase although the numbers of individual enzymes vary. Mixtures of purified endo- and exoglucanases from *T. reesei* combined in the same proportions as in culture filtrate degrade as effectively as crude culture filtrate negating the involvement of any 'special factors' for the attack of crystalline cellulose as have been suggested in the past. Amorphous cellulose is degraded by both endo- and exoglucanases separately but to attack crystalline cellulose the combined action of the two classes of enzyme seems to be necessary.

Most potential raw materials contain various polymers in addition to cellulose. Processes for the utilization of cellulosic wastes must therefore involve some pretreatments to obtain cellulose suitable for hydrolysis. Such pretreatments generally involve an initial size reduction to increase the biomass surface area that will be in contact with the second stage of enzymes, acid, solvent or steam that is being used to dissociate the cellulose, hemicelluloses and lignin. The problems with these pretreatments are that they require energy, equipment and often expensive chemicals. Furthermore, pretreatment is often accompanied by irretrievable loss of sugar and formation of undesirable by-products. Cellulase yields from the most productive microorganisms are relatively low and commercial cellulases are expensive. These problems coupled with other economic factors, in particular transport costs, currently inhibit the commercial applications of cellulase. Nevertheless, there is considerable research effort devoted to the improvement of cellulases and the processes in which they can be used and this will undoubtedly improve the economics of cellulose utilization.

β-**Glucanase**   The cell walls of the aleurone layer of barley are virtually devoid of cellulose but contain large amounts of β-glucan and hemicelluloses. The β-glucan is a linear polymer of glucose residues which are connected by 1,3-β- and 1,4-β-linkages. Various glucanases are synthesized by barley during malting and degrade the β-glucan but in brewing processes utilizing low levels of malt, the β-glucan may be hydrolysed only partially, if at all. Persistence of this viscous material leads to filtration and haze problems.

   B. amyloliquefaciens secretes 1,4-β-glucanase in addition to proteases and amylase and by appropriate adjustment of the growth conditions high yields can be obtained. The enzyme is an endoglucanase that yields small oligosaccharides from barley glucan that contain the original 1,3-β-linkages of the polysaccharide. It is stable at pH 5 and up to 50°C and has found considerable usage in cereal fermentations.

Hemicellulase   Hemicelluloses are alkali-soluble polysaccharides associated with the cellulose in plant cell walls. The principal components from higher plants are complex, substituted β-linked xylans, the chemical composition of which varies with the source of the material. The hemicellulose fraction of wood also contains 1,4-β-mannans. Interest in the use of waste plant materials for fermentation has encouraged study of hemicellulases. Xylanases are found in many fungi and endo-acting enzymes have been characterized from *Aspergillius* and *Trichoderma* species. *Apergillus niger* also produces an endomannanase. The commercial production and usage of hemicellulases is currently very limited, largely because of the difficulties in processing plant waste residues.

Invertase   Invertase hydrolyses sucrose into glucose and fructose, both of which are more soluble than sucrose. The resultant syrups find application in the food and confectionary industries. The enzyme is readily obtained from *Saccharomyces cerevisiae* strains and has been immobilized in cellulose triacetate fibres for the continuous production of invert sugar.

Lactase   Lactase (β-galactosidase) hydrolyses lactose into glucose and galactose. It is an intracellular enzyme in bacteria and yeasts but is secreted by many fungi; commercial preparations are generally manufactured from aspergilli. The potential applications for lactase, particularly in the dairy industry, are considerable. At present the high cost of the enzyme is precluding much of this technology but the availability of immobilized lactase from *A. niger* should help relieve this problem.

   Lactose has low solubility and low sweetness. Hydrolysis of the lactose in milk and dairy products avoids crystallization (for example in ice-cream) and provides sweetness with obvious commercial advantages. There are also dietary considerations in that some infants and adults (and individuals of all ages in some ethnic groups) cannot metabolize lactose due to a lack of β-galactosidase. Lactase treatment will offer readily metabolized milk and dairy products for these people. Lactose in the form of milk whey is the major by-product from cheese manufacture and is generally disposed of in the environment causing many pollution problems. Current attempts to utilize this valuable carbon supply involve its fermentation to ethanol using lactose-metabolizing yeasts, particularly *Kluyveromyces* species. Since these microorganisms are not very efficient at converting lactose to ethanol, prior treatment of the whey with immobilized lactase and fermentation with *Saccharomyces* species is being considered. Alternatively, hydrolysed whey can be concentrated into syrups for use in the food and confectionery industries.

Lipase Lipase (glycerol-ester hydrolase) is synthesized by many microorganisms, the principal sources for the commercial enzymes being *Rhizopus*, *Mucor*, *Aspergillus* and *Candida* species. These enzymes have been purified and characterized in some detail. The specificity of the enzyme varies according to its source. Thus the enzymes from *Aspergillus* and *Rhizopus* strains are similar to pancreatic lipase in that they do not attack the 2-position in the triglyceride and show low specificity for the type of fatty acid whereas *M. miehei* lipase is active on short-chain triglycerides with no marked position specificity.

Numerous applications have been proposed for these enzymes including lipase-containing washing-powders for removal of fatty materials from clothes. Currently, lipases are used in cheese manufacture for flavour development and as a digestive aid to replace pancreatic lipase which is expensive and scarce. They are also used in diagnostic enzyme kits for serum triglyceride analysis.

Pectinase Pectins are structural polysaccharides occurring within and between the cell walls of higher plants. In these locations they bind cellulose fibrils together to form a rigid matrix and they cement cells together. Pectin comprises chains of partially methyl-esterified 1,4-α-galacturonan. The demethylated molecule is known as pectic acid or polygalacturonic acid (Figure 11). Modifications of this structure occur depending on the origin of the material.

Pectinase covers a variety of different enzymes classified into two main groups: (1) pectin esterases which de-esterify pectin to produce methanol and pectic acid and (2) depolymerases which split the glycosidic bounds of the pectin substrate. Depolymerases may operate by hydrolysis of the molecule or more commonly by a *trans* elimination reaction in which cleavage of the glycosidic linkage of the molecule at C-4 is accompanied by simultaneous elimination of the hydrogen atom at C-5 to produce oligouronides which terminate in a C-4,5-unsaturated galacturonyl unit (Figure 11). Fungi tend to secrete *trans*eliminases that are specific for pectin (polymethylgalacturonate lyases) whereas the currently known

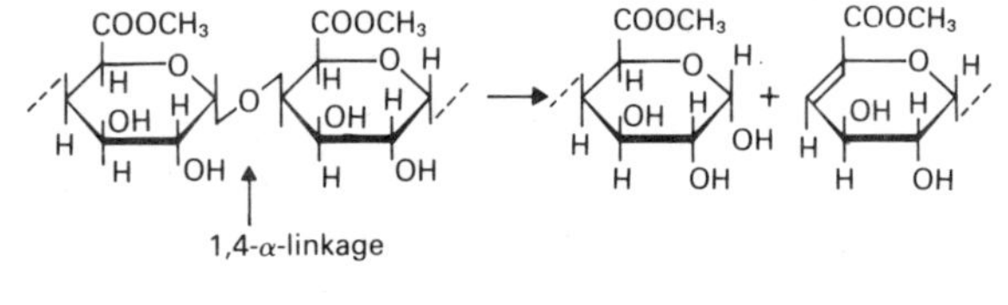

Fig. 11 Transeliminative cleavage of pectin by polymethyl-galacturonate lyase.

bacterial *trans*eliminases are most specific for pectic acid (polygalacturonate lyases) and are quite widely distributed amongst various genera, in particular plant pathogens that cause 'soft rots' of plants.

Commercial pectinase preparations are produced exclusively from *Aspergillus niger*. Not only is this a prolific source of pectin-degrading enzymes but the organism has Food and Drug Administration (FDA) approval for use in foods in the USA. The enzyme composition depends largely on the culture conditions but commercial preparations generally contain polygalacturonase, polymethylgalacturonate lyase, and pectin esterase.

Pectinases are used almost exclusively in the fruit and vegetable industries. The traditional and probably the largest application is for the clarification of apple, pear and grape juices in which pectin often gives rise to turbidity and viscosity. Such depectinization is a prerequisite for juices that are to be concentrated. By treating the pulp itself with pectinase it is possible to increase the yield as well as the quality of the juice. Pectinases are also used for wine and citrus juice clarification.

Penicillin amidase The enzyme penicillin amidase (penicillin acylase) catalyses the production of 6-aminopenicillanic acid (6-APA) from penicillin (Figure 12). This is a valuable precursor for the chemical synthesis of a diversity of substituted penicillins with various clinical uses, such as activity against certain Gram negative bacteria or resistance to β-lactamase, the main microbial defence against penicillins. Although *Penicillium* strains produce 6-APA they do so in low yield. The principal source of 6-APA is therefore from the enzymic treatment of benzylpenicillin using an immobilized penicillin amidase from *Escherichia coli*.

Immobilized enzymes

Enzyme immobilization covers many areas including the binding of enzymes to the inner surfaces of plastic tubes and membranes for use in autoanalysers and enzyme electrodes, in addition to the advantageous usage of insoluble enzymes in various industrial and technical processes. Furthermore, it has indirectly led to the

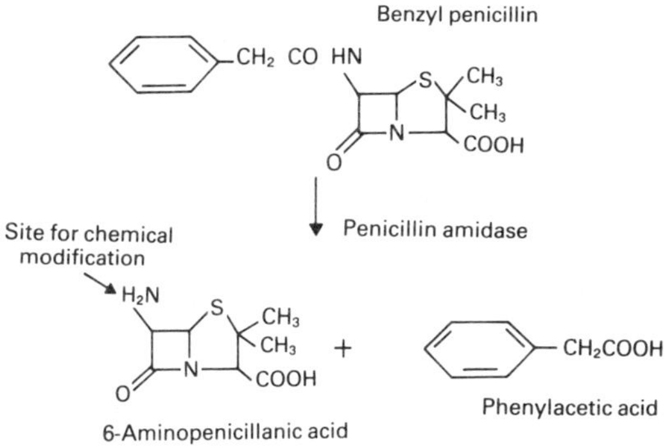

Fig. 12 Enzymic breakdown of benzylpenicillin to 6-aminopenicillanic acid by penicillin amidase.

45

development of such techniques as radio-immune assay and enzyme-linked-immunosorbent assay. In the chemical, food and pharmaceutical industries, immobilized enzymes offer several advantages over their soluble counterparts. These are (1) the enzyme can be removed from the product and re-used, thus lowering processing costs, (2) immobilized enzymes often show increased stability and prolonged activity, (3) immobilized enzymes lend themselves to continuous processes which in turn give rise to improved quality control and lower labour costs, (4) higher enzyme to substrate ratios can be achieved which lowers reaction times and (5) multienzyme systems can be considered. There are obviously some disadvantages that may be encountered. The enzyme may partially lose activity on immobilization or the cost of immobilization may be prohibitive: the initial plant investment costs will probably be higher, the process technically more complex and more skilled supervision of the process will be required. The final decision as to whether it will be advantageous to immobilize an enzyme therefore rests on many biochemical and engineering considerations in the context of each particular process.

There are essentially four strategies for enzyme immobilization.

Adsorption on an insoluble support   Immobilization of enzyme on ion exchange resin with retention of the enzyme by ionic forces has long been a simple and favoured procedure. Clinton Corn Processing Company (USA) manufacture a glucose isomerase from a streptomycete adsorbed onto DEAE-cellulose for the production of HFCS. However, the first large-scale use of enzyme immobilized in this way was the Japanese Tanade Seiyaku Company process for L-amino acid production for use in medicines and food. A chemically synthesized acyl-DL-amino acid is hydrolysed to give an L-amino acid plus unhydrolysed acyl-D-amino acid using a fungal amino acylase immobilized on DEAE-sephadex. The mixture can readily be separated by crystallization and the acyl-D-amino acid recycled.

$$\text{DL-acyl amino acid} \xrightarrow{\text{amino acylase}} \text{L-amino acid} + \text{D-acyl amino acid}$$
$$\underset{\longleftarrow\text{chemical racemization}\longrightarrow}{}$$

Inorganic carriers can also be used for enzyme immobilization, for example Corning Glass Corporation market glucose isomerase from a streptomycete adsorbed onto alumina for use in a column reactor for HFCS production.

Immobilization in the host cell   Enzyme immobilization within the cell is a popular procedure and formed the basis of the first commercial utilization of glucose isomerase. A streptomycete culture was heated for short periods at 60 to 80°C which purportedly destroyed autolysins and left cells that were stable and permeable to small molecules. These cells could then be pelleted and packed into a bed as an enzyme reactor. Imperial Chemical Industries (UK) on the other hand use a flocculating agent to fix glucose isomerase within *Arthrobacter* cells and make a cell paste which is extruded and dried in pellets for use in a column reactor. Alternatively cytoplasmic glucose isomerase can be fixed within cells of *Streptomyces* by covalent cross-linking. The most commonly used cross-linking agent is glutaraldehyde which produces a stable, three-dimensional network of proteins. This bifunctional aldehyde contains two aldehyde groups at either end of a $-(CH_2)_3-$ unit. At neutral pH the aldehyde groups will react with free amino groups. The other end can then be attached to a solid support or some other cell constituent.

The beet sugar industry uses immobilized α-galactosidase to hydrolyse the indigestible sugar raffinose into sucrose and galactose. This enzyme is fixed within pellets of the fungus *Mortierella vinaceae* which are formed under particular growth conditions and the untreated cells can be used directly as biocatalysts.

Entrapment of enzyme or cells in an insoluble matrix   Entrapment of enzymes or cells within a gelatin or polyacrylamide gel is an effective and simple procedure. Thus Gist Brocades (USA) manufacture glucose isomerase particles by mixing cells with gelatin, and retaining the enzyme by cross-linking with glutaraldehyde. Such systems can only be used for enzymes with low molecular weight substrates and products because of diffusion problems. A more complex procedure, but on similar lines, involves trapping enzyme in cellulose triacetate fibres. Typical preparations can be obtained by extrusion of an emulsion of aqueous enzyme solution with cellulose triacetate in methylene chloride. Almost 50% of the enzyme will be trapped within the resultant fibres. Because the filaments are porous, substrate and product diffuse freely across the walls of the filaments but the large enzyme molecule cannot escape. In this way glucose isomerase, amino-acylase, glucoamylase and β-galactosidase have been trapped in fibres for industrial processes.

Enzyme covalently bonded to a support   Preparations of covalently bonded enzyme generally involve rather complicated procedures. Glucose isomerase has been covalently coupled to porous glass and ceramics. Glass is mostly silica and can be reacted with an alkylamino silane to give an alkylamine glass to which enzymes can then be coupled by cross-linking. These methods are not commonly used for industrial preparations.

At present, several of these immobilization procedures seem to be competitive. In general, immobilization in cells is attractive because it is inexpensive and the enzyme retains most of its activity. Similarly, immobilization by adsorption is simple and large amounts of enzyme can be attached to provide a highly potent (activity for weight) material. In this case the cost of the carrier may be high but it can be re-used. Simpler entrapment procedures produce highly potent products but diffusion limitations often reduce the activity of these materials. Covalently-bonded enzymes tend to have low potency and low activity.

Enzyme reactors   Reactors may be broadly classified into batch and continuous systems. The batch reactor, in which the substrate is stirred in a tank with enzyme, is simple and generally used for free, soluble enzymes. No attempt is made to recover the enzyme from the product. The enzyme is generally heat inactivated. When immobilized enzymes are used in a batch reactor, the enzyme is recovered by filtration or centrifugation. These procedures often result in appreciable loss and inactivation of expensive enzyme and consequently such reactors have limited application for immobilized enzyme processes.

Continuous flow reactors may be of the packed-bed or stirred-tank configuration. The fluidized-bend is a hybrid of these two types (Figure 13). Immobilized enzyme in the form of spheres, chips, disks, beads, pellets or fibres can be readily packed into a column to form a packed-bed reactor. Substrate is fed into the reactor through the bed of immobilized enzyme. If the fluid velocity profile is

Extracellular Enzymes

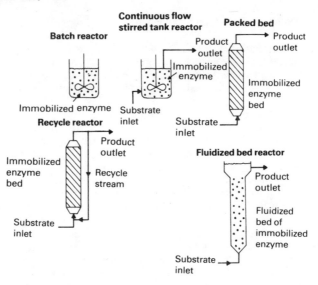

Fig. 13 Types of enzyme reactors (from Vieth *et al.*, 1976).

perfectly flat over the cross-section, the reactor is said to operate as a plug-flow reactor since the fluid is visualized as moving through the reactor in a plug-like fashion. The contents of this reactor are therefore heterogeneous with maximum substrate concentration on entry and maximum product concentration on exit. Consequently these reactors are especially suited to enzymes that undergo product inhibition. Most of the published immobilized-enzyme reactor studies deal with packed-bed reactors and they form the basis of industrial usage of glucose isomerase.

In an ideal, continuous-flow, stirred-tank reactor, the contents are in a steady state, and the system resembles the chemostat used for continuous culture of microorganisms. These systems lend themselves to easy control of temperature and pH, but the average reaction rate is lower than it would be in a tubular reactor except for the rare instances when an enzyme is substrate inhibited. Immobilized enzyme must be retained within the reactor and this is readily achieved by providing a filter at the outlet.

In a fluidized-bed reactor, the substrate flows upward through the immobilized enzyme sufficiently rapidly to lift the particles but not so fast that particles are swept away in the exit stream. The fluid flow pattern therefore lies between complete back mixing (as in the continuous stirred-tank reactor) and no back-mixing (as in the plug-flow reactor). Fluidized bed reactors have found some application, particularly for the processing of viscous solutions such as gelatinized starch.

Summary

Carbohydrases and proteases comprise 90% of the industrial enzyme market. Of the former, starch hydrolysing enzymes predominate, in particular α-amylase from *Bacillus* strains. This enzyme hydrolyses the 1,4-α-bonds of starch in a random, endo-fashion and causes a rapid reduction in the size of the substrate. Their main use is to 'liquefy' starch at high temperature. Glucoamylase is an exo-attacking enzyme secreted by various moulds that removes glucose residues from 1,4-α-glucans. In conjunction with α-amylase, it will reduce starch to glucose which can be converted to fructose in high yield by the enzyme glucose isomerase. Since fructose tastes very sweet, high fructose syrups are replacing sucrose in many foods.

Proteases may be assigned to three categories, those with a serine residue at the active site, those with a metal ion requirement and those with an acid pH optimum. The serine proteases from some bacilli are temperature stable and have an alkaline pH optimum which makes them ideal for inclusion in household washing powders. The metalloproteases, also from bacilli, are optimally active at neutral pH and are used in the food industry. Acid proteases are secreted by various fungi, and those from *Mucor* and *Endothea* strains are replacing calf rennet for the manufacture of cheese. The remaining 10% of the market is occupied by several enzymes that have either limited application or large potential markets that are currently uneconomical. Fungal pectinase is used to clarify fruit juices and wines where pectin hazes may be problematical. Several fungi also secrete β-galactosidase which has potential for the hydrolysis of lactose in milk whey, a major by-product from cheese manufacture, into glucose and galactose. However, the most attractive potential market is the production of glucose from waste plant material using fungal cellulase, but an economic process has yet to be developed.

By immobilizing enzymes on solid supports, it is possible to recover the enzyme and re-use it. Alternatively, the immobilized enzyme can be packed into a column or similar reactor and used in a continuous process. In this way more expensive, intracellular enzymes such as glucose isomerase can be considered for industrial purposes.

References

ANTRIM, R. L., COLILLA, W. and SCHNYDER, B. J. (1979). Glucose isomerase production of high-fructose syrups. In: *Applied Biochemistry and Bioengineering*. Vol. 2, *Enzyme Technology*, pp. 98–115. Edited by L. B. Wingard, Jr., E. Katchalski-Katzir and L. Goldstein. Academic Press, London and New York.

AUNSTRUP, K. (1980). Proteinases. In: *Microbial Enzymes and Bioconversions*, pp. 50–114. Edited by A. H. Rose. Academic Press, London and New York.

AVERINGOS, G. C. and WANG, D. I. C. (1979). Direct conversion of cellulosics to ethanol. *Annual Reports on Fermentation Processes* 4; 165–91.

ENARI, T.-M. FOGARTY, W. M. and KELLY, C. T. (1983). In: *Microbial Enzymes and Biotechnology*. Edited by W. M. Fogarty. Applied Science, London. Chapters entitled Microbial amylases, Pectic enzymes and Microbial cellulases.

FUKUI, S. and TANAKA, A. (1982). Immobilized microbial cells. *Annual Review of Microbiology* 36; 145–72.

KATCHALSKI-KATZIR, E. and FREEMAN, A. (1982). Enzyme engineering reaching maturity. *Trends in Biochemical Sciences* 7; 427–30.

PHILLIPS, R. R. (1980). Approach of the enzyme manufacturer to enzyme developments and markets. In: *Enzymes: the Interface Between Technology and Economics*, pp. 11–16. Edited by J. P. Danely and B. Wolnack. Marcel Dekker Inc., New York.

Extracellular Enzymes

TAYLOR, M. J. and RICHARDSON, T. (1979). Applications of microbial enzymes in food systems and in biotechnology. *Advances in Applied Microbiology* 25; 7–36.

VIETH, W. R., VENKATASUBRAMANIAN, K., CONSTANTINIDES, A. and DAVIDSON, B (1976). Design and analysis of immobilized-enzyme flow reactors. In: *Applied Biochemistry and Bioengineering*. Vol. 1, *Immobilized Enzyme Principles,* pp. 221–327. Edited by L. B. Wingard, Jr., E. Katchalski-Katzir and L. Goldstein. Academic Press, London and New York.

5 Strain selection and improvement

Any biological process resulting in a commercial product must be developed with regard to two related considerations. First, there must be a market for the product or the new or improved process and second, it must be possible to manufacture the product or market the process at an economic price. Obviously, if an enzyme is sold at a high price the profit margin for the manufacturer will be acceptable, but alternative and cheaper processes will be used by the customer. Since there are many more enzymes available than commercial processes in which they can be used, it would seem prudent to assess the market demand for new or improved existing processes and then search for an appropriate enzyme. Having found a potential market and suitable enzyme source, improvement of enzyme yield is the major factor that will decide if the manufacture of the enzyme, and consequent implementation of the new process, will be economically viable.

Strain isolation

Sources of microorganisms Extracellular enzymes must function in the immediate environment of the microorganism if they are to be of any value to the host. Consequently, thermophilic bacteria might be expected to secrete temperature stable enzymes, and alkalophiles, enzymes that are optimally active at high pH. This is generally found to be the case, for example *Bacillus acidocaldarius* is a thermophilic, acidophilic bacterium that inhabits acid, hot springs. The organism grows well at 50 to 80°C and pH 2 to 6 and the α-amylase that it secretes is optimally active at pH 2 to 4.5 and 70°C. The optima for the α-amylase from the mesophile *B. amyloliquefaciens* on the other hand are 65°C and pH 6, the enzyme retaining only 15% of the activity at pH 3.5. There are exceptions to this rule, however, and the α-amylase from the mesophile *B. licheniformis*, which hydrolyses starch under pressure at 110°C, is a prime example.

Having defined a process and its operating conditions, the microbiologist will have some idea of the nature of enzyme and hence the organism required. This will determine the sites to be sampled. Thermophiles will predominate in hot springs, self-heating composts, desert soils and similar habitats, but will also be recovered from typical garden soils. Similarly alkalophiles will predominate in areas rich in lime, for example effluent from cement manufacturing plants, but also occur widely in neutral soils. It is advisable to sample sites that are rich in the particular nutrient of interest; cellulase and hemicellulase secreting organisms will be common in forest litters and composts, pectinase producing organisms in decomposing fruits and vegetables.

Although the environment offers an enormous diversity of microorganisms, established culture collections should also be considered since they usually offer a wide selection of strains from geographically dispersed sites and often include rare strains that are seldom encountered in the environment. Most culture collection catalogues give lists of enzyme-producing strains held in the collection.

Screening for extracellular enzymes When bacteria or filamentous fungi are grown on a solid medium, extracellular enzymes diffuse into the agar gel around the colony. If substrate for the enzyme is incorporated into the medium, the presence of enzyme activity can be revealed as a zone of degraded substrate. In the case of cell bound enzymes, it may be necessary to remove the colony from the plate to reveal the patch of substrate hydrolysis. This technique forms the basis of several simple and rapid screening procedures for extracellular enzymes (Table 3). With insoluble substrates (for example chitin or purified cell walls from some other organism), the material is incorporated into the medium as the sole or a supplementary carbon source; clearing zones around productive colonies are apparent after incubation. For soluble substrates, it is necessary to either precipitate or stain the polymer to reveal clearing zones. This often has the disadvantage that the colonies are killed by the added reagent and it is necessary to make an initial replica plate of the colonies.

An alternative procedure is to couple the substrate to a dye; remazol brilliant blue and azure are commonly used. Starch azure takes the form of insoluble blue granules which are hydrolysed by α-amylase but the dye molecules inhibit exoattack by β-amylase. When incorporated into an agar medium and enzymically hydrolysed, the low molecular weight α-limit dextrins diffuse away from the colony to leave a pale blue halo. This test is therefore specific for α-amylase. Similarly, RNA complexed with acridine orange appears green under UV illumination and hydrolysis is revealed as zones. This method is more sensitive than precipitating the RNA with HC1 and the microorganism is not killed.

Extracellular and cell bound enzymes that hydrolyse low molecular weight substrates require different screening procedures. Chromogenic substrates (for example p-nitrophenyl-α-glucoside for α-glucosidase) that release coloured products can be used, or the products of the enzyme reaction can be detected using added reagents. Since these small molecules are readily diffusible, it is generally advisable to incorporate the reagents in soft agar, overlay developed colonies, and inspect for enzyme activity within a short time.

Such screening procedures allow the rapid examination of many hundreds of colonies but, from heterogeneous starting material, it may be necessary first to enrich the inoculum material.

Enrichment The concept of enrichment is to provide conditions that promote the growth of organisms with the desired characteristics such that the relative numbers of these organisms increase. This can be achieved in closed systems by operating successive batch cultures in which, for example, starch is provided as sole or principal carbon source. Those organisms best suited for starch metabolism under the conditions imposed will ultimately predominate and amylase producers can be detected by plating onto starch agar. A very strong selection pressure and hence enrichment can be obtained from continuous culture. By operating a chemostat with starch as the limiting carbon source and soil as the inoculum, a single organism, or often a community of several organisms, will be selected that use starch most efficiently under the conditions used. All other organisms from the original material will be 'washed out' of the growth vessel.

These enrichments are ideal for intracellular enzymes but with extracellular enzymes the products of the enzyme's action benefit all organisms in the culture. It is therefore important to examine batch culture enrichments early in the growth cycle before the degradation products permit growth of competing organisms. In

Table 3 Media for the detection of extracellular and cell-bound enzymes

Enzyme	Substrate	Reagents	Comments
Polysaccharases			
Amylase	Starch/glycogen	Iodine solution	Clear (α-amylase) or red (β-amylase) zones against stained background.
α-Amylase	Starch azure	—	Pale blue halo.
Cellulase	Cellulose azure	—	Pale blue halo.
	Filter paper		Degradation of filter paper.
Chitinase	Chitin	—	Clear zone in insoluble chitin.
α-Glucosidase	p-Nitrophenyl-α-D-glucoside	—	Yellow colonies and zones.
α-Glucanase	1,3-α-and/or 1,6-α-glucan	—	Zones in insoluble dextran.
1,3-β-Glucanase	Pachyman	Aniline blue	Substrate and reagent form complex, pale blue zones created.
1,6-β-Glucanase	Pustulan	Ethanol	Clear zones in precipitated substrates.
Petate lyase	Apple pectin or polygalacturonic acid	Hexadecyl-methylammonium bromide	Clear zones in precipitated substrate.
Pullulanase	Pullulan	Ethanol or acetone	Clear zones in substrate after 16 h.
Xylanase	Xylan	Ethanol	Clear zone in precipitated substrate after 16 h.
Cell wall lytic enzymes	Purified cell walls		Zones of clearing.
Proteolytic enzymes			
Proteases	Skimmed milk	—	Zones of clearing.
	Gelatin	Saturated ammonium sulphate	Zones of clearing in precipitated substrate.
	Hide powder azure	—	Pale blue hydrolysis zones.
Nucleases			
Deoxyribo-nuclease	DNA	HCl	Zone of clearing in precipitated DNA.
	DNA	Methylgreen	Autoclave methyl green with medium. Pink zones around colonies.
Ribonuclease	RNA	HCl	As for DNA.
	RNA	Acridine orange	Viewed under UV light dark zones in fluorescent green background.
Lipase	Tween		Tween is included in a suitable medium. Opaque haloes indicate lipolytic activity.
Penicillinase (β-lactamase)	Penicillin	Polyvinyl alcohol/ I_2 solution	Polyvinyl alcohol in medium reacts with I_2 giving dark blue colour. Penicillinoic acid from β-lactamase reaction clears the blue colour.

the chemostat, organisms that secrete enzymes into the culture are seldom selected; instead bacteria with tightly cell-bound enzymes tend to predominate. This is because there is no competitive advantage for an organism that secretes an enzyme into the culture fluid thereby benefitting all the organisms in the growth vessel.

Desired properties of commercial strains   In addition to secreting high yields of the desired enzyme, good commercial strains must also comply with several other criteria. The organism should be robust and grow readily in large scale fermenters. It should produce a minimum of other enzymes (unless a mixture of enzymes is desired) and should not secrete toxic metabolites such as allergens or antibiotics. The organism should not be pathogenic and, if the enzyme is to be used in foods, it is preferable that the organism be passed as safe by the appropriate national bodies.

Increasing enzyme yield

Mutation   Classical mutation and selection programmes simply involved screening mutated cultures for high yields of product. They were very successful in the antibiotic industry during the 1950s and 1960s in the absence of fundamental knowledge concerning the regulation of secondary metabolism. From the screening of hundreds of thousands of strains, superior antibiotic-producing variants were selected that formed the basis of the antibiotic industry.

Many mutagens are currently available for this type of programme and, since they act at different sites and in different ways, it is advisable to make use of the variety. Thus a high-yielding mutant might be generated using ultra-violet light and then this organism mutated with an alkylating agent to obtain a further-improved strain. During mutation, a moderate mutation rate is advisable since an excessive mutation rate leads to multiple mutations, and the introduction of several deleterious mutations for each beneficial one.

Chemical mutagens are dangerous to handle (by their very nature they are carcinogenic) and some microorganisms are highly resistant to mutation by these agents. For these reasons, transposons have been widely used as genetic mutagens. Transposons are segments of DNA that can insert at many loci within genomes. They encode proteins that are necessary for the transposition and some other determinant, for example antibiotic resistance. By their insertion they can mutate genomes, often giving rise to highly polar mutations. Not all transposons are equally useful as mutagens, since some do not insert into chromosomes very readily. The kamamycin resistance transposons Tn5 is useful since it transposes readily with little target specificity. A number of strategies have been devised for transposon mutagenesis: a straightforward procedure is to introduce the transposon into a host cell population on a piece of DNA that cannot replicate in the recipient and to select for the antibiotic resistance of the transposon (Starlinger, 1980).

Having mutated a culture, screening for improved enzyme yield is often performed using media such as those described in Table 3. It cannot be over emphasized that the increase in zone size in such media is directly proportional to the *logarithm* of the enzyme yield. This method is therefore not very useful nor accurate for detecting small increases in yield from existing hyper-yielding strains.

For example, an increase in enzyme yield from 4000 to 6000 units per ml represents a very valuable 50% increase in enzyme production but the zone size around a colony would only increase from 3.6 to 3.8 cm which could not be detected. Consequently mutants should always be checked by assaying culture fluid from standardized broth cultures, a very time-consuming exercise. Nevertheless, starting with low-yielding *B. subtilis* strains, mutants that secrete about 5-fold more amylase have been detected in this way.

Selection for hyperproduction Although there are no direct selection procedures for mutants that secrete high yields of extracellular enzymes, the inefficiencies of screening have encouraged the use of several indirect selection strategies. These generally involve selecting for readily-selectable markers in the hope that such mutants will be pleiotropically affected in the regulation of extracellular enzyme synthesis or secretion. The stages at which extracellular enzyme synthesis is regulated in prokaryotes are probably: (1) RNA polymerase binding and initiation, (2) induction or derepression of transcription, (3) catabolite repression of transcription, (4) ribosome binding and mRNA stability, and (5) passage of the enzyme through the membrane and wall with accompanied processing. By selecting for mutants with modifications in these operations, organisms that secrete high yields of extracellular enzyme can be isolated.

RNA polymerase mutants can be readily isolated by selecting for resistance to rifampicin or streptolydigin. Rifampicin binds to the β subunit of the enzyme and inhibits transcription by preventing translocation immediately after binding and the formation of the first phosphodiester bond of the mRNA. Streptolydigin inhibits mRNA elongation, and mutations to resistance in *E. coli* and *B. subtilis* are located in the β and β' subunits of the enzyme, respectively. Rifampicin and streptolydigin resistant mutants of *B. subtilis* are often pleiotropic, sporulate at reduced frequency or not at all and are affected in extracellular enzyme synthesis. This may be due to altered recognition of the extracellular enzyme gene promoters by the mutated RNA polymerase. However, whenever sporulation is affected, it is likely that the complex phenotype results from disruption of the finely tuned, physiological network essential for sporulation.

Induction or derepression controls the expression of many extracellular enzyme operons, and removal of inducer requirement can lead to high enzyme yields and remove the expense of adding inducers to industrial fermentations. The use of substrates for enzymes that do not themselves induce enzyme synthesis provides strong selection pressure for constitutive synthesis. Thus phenyl-β-D-galactoside can be used to select for constitutive synthesis of β-galactosidase in *E. coli* and an example of industrial significance concerns glucose isomerase synthesis in *Streptomyces phaeochromogenes*. This enzyme is used for the metabolism of both xylose and lyxose, but lyxose does not induce its synthesis. By selecting for spores that will germinate on lyxose, mutants constitutive for glucose isomerase synthesis were readily obtained.

Mutants that are insensitive to catabolite repression are valuable in several respects. They often synthesize and secrete extracellular enzymes earlier in the batch culture growth cycle than wild type strains thus improving production efficiency. They almost invariably secrete higher levels of extracellular enzymes, and they allow the use of cheap fermentation media often containing large amounts of glucose. A simple screen to obtain such mutants employs a high concentration of glucose, glycerol or some other rapidly metabolized carbon

55

source in the screening agar. Mutants able to produce normal degradation zones under these conditions are at least partially resistant to catabolite repression. Antimetabolites have also been used successfully for selection of catabolite derepressed mutants. Strains of *Trichoderma* have been isolated that synthesize β-glucosidase in the presence of glucose, by selecting for growth on cellobiose and the toxic glucose analogue, 2-deoxyglucose. The parent strain is inhibited by the 2-deoxyglucose; mutants unable to utilize glucose form large colonies by growing on the cellobiose. A third approach is to select for sporulation in the presence of glucose by pasteurizing glucose-grown cultures. Such mutants are catabolite derepressed for sporulation and also produce an increased yield of serine protease due to a mutation in the *catA* locus.

With regard to translational control, there are three possibilities for modification; increased mRNA stability, improved translation initiation sites, and faster chain elongation. The rate of chain elongation is largely invariable, but there is some evidence that polypeptide chain initiation frequency is message-dependent and therefore amenable to improvement by mutation. Certainly some form of translational control is implicated in *Bacillus*, since various mutations that offer resistance to ribosome inhibitors (e.g. kanamycin and erythromycin resistance) also affect sporulation. Such mutants sometimes secrete very low or high levels of extracellular enzymes but, as with rifampicin resistance, this is probably due to an upset in the control network for sporulation rather than a specific effect on extracellular enzyme synthesis.

Mutations in the secretory process have been particularly useful for increasing enzyme yields, although the molecular details are seldom understood. Mutants of *B. subtilis* that hypersecrete extracellular enzymes have been obtained by selection for resistance to various antibiotics that affect cell wall structure such as cycloserine, ampicillin, novobiocin and tunicamycin. Similarly, modification of the membrane is a prime target, and polyene (kabasidine) resistant mutants of *Fusarium* and *Trichoderma* species have been found to hypersecrete extracellular enzymes. It should be noted that in few, if any, instances has the cell surface modification been directly linked with the enzyme hypersecretion phenotype.

Gene transfer

During a strain improvement programme involving sequential mutation, several lineages of high-yielding organisms may be developed. Since it is unlikely that the same mutations will be responsible for the excessive production of enzyme in each strain, the recombination of these genotypes by genetic exchange can give rise to exceptionally high yields. Figure 14 shows part of an idealized strain development programme. If each improvement in yield results from a single, different mutation, by crossing strains 5 and 10, over 1000 different genotypes could be generated. In many instances, the phenotypic effects of the mutations will be additive, and very high enzyme yields will result from the recombinants. Alternatively, individual beneficial mutations can be accumulated in a single strain during several transfers.

Classical gene transfer techniques can be used to obtain the recombinants. In Gram negative bacteria, conjugation using either F-plasmid mobilization of the chromosome in *E. coli* and related bacteria, or some other transferable plasmid with chromosome mobilizing ability for other genera could be used. Conjugation

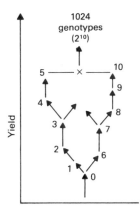

Fig. 14 Part of an idealized strain improvement programme. If it is assumed that each yield-enhancing step, after divergence from strain 0, represents mutation in a different gene, 2^{10} genotypes are produced by crossing strains 5 and 10. (From Hopwood, 1979.)

is also available for *Streptomyces* and some other Gram positive bacteria. However, the most successful development programme for increasing enzyme yield that used this approach involved transformation in *B. subtilis*.

Transformation *Bacillus subtilis* is naturally transformable; that is under certain growth conditions competent cells will bind exogenous DNA that subsequently enters the cell. If the foreign DNA is largely homologous with the host DNA, recombination occurs between the two molecules and stable transformants are produced. Maruo and his co-workers have exploited this process to construct strains of *B. subtilis* that secrete vast amounts of α-amylase.

B. subtilis Marburg, the wild type strain, produces about 10 units ml^{-1} amylase under the influence of the regulatory gene *amyR1*. The more active regulatory gene *amyR3* from a high-yielding mutant generates five times this amount of amylase and the gene *pap*, probably responsible for an alteration in the secretion process, promotes a threefold increase in the yield of both amylase and protease. When individually introduced into *B. subtilis* Marburg by transformation, these loci gave rise to the expected fivefold and threefold increases in amylase yield. When both were combined in the same strain the effect was multiplied and a 14-fold increase in amylase yield was observed (Table 4). More high-yielding genotypes were generated in the parent Marburg strain by mutation and selection for resistance to various antibiotics coupled with increased hydrolysis zones in starch-nutrient agar. For example, from about 5000 cycloserine resistant mutants, some 200 showed increased α-amylase and/or protease activity and from these, strain C-108 was selected which produced 5 times more α-amylase and 80 times more protease than the wild type. When DNA was isolated from these mutants and incorporated into the single strain by transformation, the amylase productivity was increased with each mutation until a strain that secreted about 15 000 units ml^{-1} of α-amylase (a 1500-fold increase over the original strain) was developed (Table 4). Thus regulatory mutations that individually have a relatively small effect on enzyme yield may act synergistically when introduced in a single strain and greatly enhance product formation.

Table 4 Effect of combining regulatory genes for α-amylase synthesis in *Bacillus subtilis*

amyR1[1]	amyR3[2]	tmrA7	amyS	Pap	C108	N-26	Relative α-amylase yield
+							1
+				+			2
	+						5
	+			+			14
	+	+					20
	+		+	+			20
	+	+	+	+			120
	+	+	+	+	+		700
	+	+	+	+	+	+	1500

Data from Yamane and Maruo, 1980.
1. Strain containing *amyR1* only is the type strain *B. subtilis* Marburg.
2. *amyR3* is derived from *B. subtilis* var. *amylosacchariticus* and is pheotypically identical to the mutant locus *amyR2*.

Protoplast fusion   Although any gene transfer technique can be used to construct such high yielding strains, many bacterial and fungal groups lack a suitable system. For these organisms protoplast fusion may hold the answer. If the cell wall is removed from Gram positive bacteria or fungi using enzymes in an isotonic medium, spherical protoplasts result. These osmotically sensitive structures are bounded by the cytoplasmic membrane and, in the presence of polyethylene glycol and Ca^{2+}, will dehydrate and agglutinate. Fusion then occurs between small regions of membrane in contact, cytoplasmic bridges are formed which enlarge, and the protoplasts fuse. In the fused product, if the two chromosomes are largely homologous, recombination will occur. By incubating on appropriate media, the protoplasts grow a cell wall and revert to the normal microbial form. Protoplast fusion has been demonstrated in various Gram positive bacteria including bacilli, *Brevibacterium flavum*, streptomycetes and *Micromonospora* species. Yeasts and filamentous fungi (penicillia and aspergilli) are also amenable to protoplast fusion, but Gram negative bacteria are difficult to manipulate in this way.

Protoplast fusion offers the unique opportunity of bringing together two complete genomes and a very high frequency of recombinants can be obtained.

Gene cloning

Developments in genetic engineering allowing the stable maintenance and expression of foreign genes in certain microorganisms offer several opportunities to the applied microbiologist interested in industrial enzymes. Increases in enzyme yield have been obtained by cloning the α-amylase gene from *B. amyloliquefaciens* into a multicopy plasmid, thereby increasing the number of gene copies 50-fold.

However, more sophisticated procedures have been developed to maximize expression of cloned genes. One particularly successful procedure [Guarente *et al.*

1980] concentrates on placing a *lac* promoter at several different positions in front of the gene in question to discover the position which provides the maximal translation. In the case of eukaryotic genes, a hybrid ribosome binding site comprising the Shine-Dalgarno sequence from the promoter fragment and the initial ATG from the gene would probably be optimal. For genes from prokaryotes, particularly those from Gram negative bacteria, such a hybrid would probably be unnecessary. In brief, to obtain the optimal position for the promoter, the foreign gene is inserted into a restriction site in a plasmid. The plasmid is then cleaved at a second site and DNA is trimmed away using appropriate exonucleases. The plasmid is then cleaved with a third endonuclease, for example *Pst* and the promoter fragment which has also been cleaved with *Pst* is ligated into the plasmid at the *Pst* site while its other end is blunt-end ligated to the exonuclease treated DNA (Figure 15). The distance between the promoter and the cloned gene will therefore depend on the amount of exonuclease digestion that has taken place and clones are examined until one that displays maximal expression is found. In this way, a plasmid that directs the synthesis of 10 000 to 15 000 molecules per *E. coli* cell of rabbit β-globin has been constructed. Other procedures which use synthetic DNA sequences attached to the gene to direct initiation of translation of foreign genes in *E. coli*, such as that for human growth hormone, have also proved highly successful.

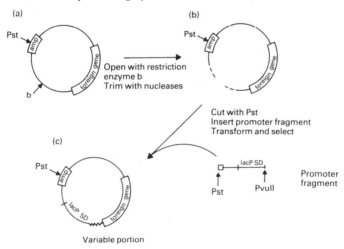

Fig. 15 Generalized procedure to maximize expression of any cloned gene in *Escherichia coli*. (Adapted from Guarente et al., 1980.)
(a) A gene is inserted into the restriction site of a particular plasmid. It can be fused to a *lac z* gene at this point so that a non-functional gene product can be assayed as β-galactosidase.
(b) The plasmid is cut with restriction enzyme *b* and the ends trimmed with exonucleases such as exonuclease III and SI.
(c) The plasmid is cut at the *Pst* site in the *amp* gene and ligated with an excess of the portable promoter fragment that contains the *lac* promoter and Shine-Dalgarno (SD) sequence. One end of the promoter fragment contains a complementary *Pst* site, the other has a flush *Pvu* II site that is blunt-end ligated to the exonuclease-treated plasmid. The resultant plasmid has a variable portion of DNA between the SD sequence of the *lac* promoter and the foreign gene.

A second benefit can be envisaged in the production of enzyme mixtures from a single host. For example, α-amylase cannot fully degrade starch because it cannot attack the 1,6-α-branch points of amylopectin. By introducing a gene for pullulanase into *B. amyloliquefaciens*, an enzyme mixture could be produced at little or no extra cost that would completely degrade starch into linear oligosaccharides. Similarly tailored combinations of cellulolytic and hemicellulolytic enzymes for the hydrolysis of agricultural wastes could be produced from single organisms. Thirdly, a novel and commercially exploitable enzyme may be produced by a suspected or known pathogen or an organism that secretes undesirable antibiotics. Toxicity testing of such strains is prohibitively expensive, particularly if the product is to be used in the food industry. If the gene for the enzyme could be transferred into a host approved for food usage (e.g. *B. subtilis*), it may be possible to reduce dramatically the amount of testing required and encourage the introduction of new enzymes and processes.

Perhaps the most exciting possibility in this area is the development of genetically engineered bacteria or yeasts that secrete proteins of commercial importance and this has provided much of the impetus for recent developments in protein secretion. One approach is to use 'secretion vectors' for the gene cloning, in which restriction endonuclease sites are situated downstream from the promoter and signal sequence of an exported protein. When a cloned gene is inserted into this site and expressed, it will bear an NH_2-terminal signal sequence that will initiate export. Depending on the sequence of the cloned gene, secretion may be achieved. This approach has been partially successful but difficulties have been encountered in getting the product beyond the periplasm of *E. coli*, and it is important to note that at present only proteins that are secreted by their host cell have been engineered such that they are secreted by the host into which they have been cloned. However secretion of eukaryotic secretory proteins such as interferon and insulin has been achieved in *E. coli* and *B. subtilis*. It may be that a Gram positive cell such as *B. subtilis* with its simpler cell wall will be more successful. In this context, the β-lactamase gene of *B. licheniformis* is being developed for secretion of eukaryotic gene products (Chang *et al.*, 1982). A plasmid has been constructed that contains the cloned cDNA of the human preproinsulin gene fused to the promoter and signal sequence of the penicillinase structural gene *penP* (Figure 16). Immunoreactive material corresponding to human insulin C-peptide could be recovered from cell extracts and culture supernatant of *B. subtilis* cultures containing this plasmid indicating that the fused protein was indeed secreted.

It would appear that the molecular details of secretion have been highly conserved throughout evolution and eukaryotic signal sequences can function in *E. coli*. An alternative strategy for secreting proteins can therefore be envisaged.

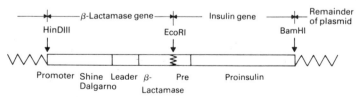

Fig. 16 Detail of the fusion gene for the expression and secretion of human proinsulin (Adapted from Chang *et al.*, 1982.)

That is, foreign genes for proteins with their own signal sequences may be cloned into a suitable host and the products secreted. As mentioned above, the gene for rat preproinsulin has been cloned into *E. coli*. The eukaryotic signal sequence directed translocation of the protein across the cytoplasmic membrane and the precursor protein was correctly processed to yield proinsulin. Similarly, a bacterial β-lactamase gene has been expressed and the precursor product processed in *Saccharomyces cerevisiae*.

Summary

Screening procedures for extracellular enzymes involve incorporating the enzyme substrate in a solidified medium. Enzyme activity is revealed as degradation zones around the colony after precipitation or staining of the substrate as appropriate. By sampling organisms from exotic locations, enzymes capable of operating at extremes of temperature and pH can often be obtained. Enrichment by successive rounds of growth in liquid media may assist the recovery of rare organisms from a sample. Having isolated the appropriate producer-organism, yield improvement is usually essential for an economical production process. Mutation and screening programmes are labour-intensive and tedious but have been successful. Selection for an increased yield of extracellular enzyme is rarely possible, but selection for some other mutation that may also increase enzyme secretion has been successful. Thus some tunicamycin and cycloserine resistant mutants of *B. subtilis* secrete increased amounts of α-amylase. Very high enzyme yields can be obtained by combining several mutations in the one host using classical gene transfer techniques such as DNA-mediated transformation. Finally, gene cloning is in its infancy with these bacteria, but the prospect is that any protein or enzyme may be made extracellular. In cases where this is particularly difficult, genetic engineering techniques will enable proteins to be made at such high levels intracellularly that recovery and purification will become commercially feasible.

References

CHANG, S., GRAY, O., HO, D., KROYER, J., CHANG, S.-Y., McLAUGHLIN, J. and MARK, D. (1982). Expression of eukaryotic genes in *B. subtilis* using signals of *penP*. In: Molecular Cloning and Gene Regulation in Bacilli, pp. 159–69. Edited by A. T. Ganesan, S. Chang and J. A. Hoch. Academic Press, New York and London.

EVELEIGH, D. E. and MONTENENCOURT, B. S. (1979). Increasing yields of extracellular enzymes. *Advances in Applied Microbiology* 25: 57–74.

FIRENCZY, L. (1981). Microbial protoplast fusion. *Symposium of the Society for General Microbiology* 31: 1–34.

GUARENTE, L., LAUER, G., ROBERTS, T. M. and PTASHNE, M. (1980). Improved methods for maximizing expression of a cloned gene: a bacterium that synthesizes rabbit β-globin. *Cell* 20: 543–53.

HOPWOOD, D. A. (1979). The many faces of recombination. In: Genetics of Industrial Microorganisms, pp. 1–9. Edited by O. K. Sebek and A. I. Laskin. American Society for Microbiology, Washington D.C.

MALIK, V. P. (1979). Genetics of applied microbiology. *Advances in Genetics* 20: 38–126.

Extracellular Enzymes

PALVA, I. (1982). Molecular cloning of alpha amylase gene from *Bacillus amyloliquefaciens* and its expression in *B. subtilis. Gene* 19: 81–7.

QUEEN, S. W. and BALTZ, R. H. (1979). Genetics of industrial microbes. *Annual Reports on Fermentation Processes* 3: 5–45.

SHERRAT, D. (1981). Gene manipulation *in vivo. Symposium of the Society for General Microbiology* 31: 35–47.

STARLINGER, P. (1980). IS elements and transposons *Plasmid* 3: 241–9.

YAMANE, K. and MARUO, B. (1980). *B. subtilis* α-amylase genes. In: *Molecular Breeding and Genetics of Applied Microorganisms,* pp. 117–23. Edited by K. Sakaguchi and M. Ouishi. Academic Press, New York and London.

YOUNG, F. E. (1980). Impact of cloning in *Bacillus subtilis* on fundamental and industrial microbiology. *Journal of General Microbiology* 119: 1–15.

6 Commercial production

Industrial scale culture of microorganisms is generally referred to as fermentation. This is somewhat misleading since the culture conditions are often highly aerobic and the microorganisms are using molecular oxygen as the final electron acceptor (aerobic respiration) rather than organic compounds (fermentation). Nevertheless, the term persists for all industrial scale production from microorganisms whether aerobic or strictly anaerobic conditions are employed. The growth of microorganisms (fermentations), to obtain enzymes, can be conveniently divided into two categories depending on the growth vessel used. Traditionally, microorganisms were grown as surface cultures on solid or semi-solid media in trays kept in a constant temperature room. For example, in the production of fungal amylase, a mixture of bran and starch is spread out thinly in trays and moistened with a mineral solution containing some hydrochloric acid. The trays are autoclaved, which hydrolyses some starch, cooled, and inoculated with a spore suspension of *Aspergillus oryzae*. After several days at 30°C a sporing mass of mycelium develops which may be dried and ground to give a crude amylase preparation or extracted to provide a semi-purified amylase. Although this process is still widely used in Japan for fungal enzymes, it is avoided in Western countries because it occupies too much space, contamination is difficult to avoid and product yield is often low.

Submerged culture methods dominate the industry today because modern methods of process control can be easily adapted to the plant. Moreover, yields are generally higher and the risk of contamination low. For this the inoculated culture is grown in a large stainless steel vessel with aeration and stirring. The vessel is then emptied, the product extracted and the process repeated. Most industrial fermentations are batch processes, but continuous processes in which medium is continuously added to a fermenter and culture removed are gradually being introduced on a commercial scale. The production methods described in this chapter will refer exclusively to submerged fermentations.

Scale-up

Before describing the process itself, it is important to consider the problems involved in taking a laboratory-scale process into industry. Difficulties are introduced first from the nature of the process itself. Consequently, large-scale versions of a process may use materials and techniques quite different from those used in the original laboratory-scale process.

'Bulk' problems associated with large-scale operation are typified by the long time periods required for sterilization. Absolute or virtual sterilization is easily achieved in the laboratory by autoclaving for short periods. On an industrial scale, one spore in a batch of medium (10,000 to 50,000 litres) is theoretically capable of causing contamination. Thus long time periods are needed to sterilize the medium (including heating and cooling) and medium deterioration is inevitable. Problems may also arise from the different materials in use; commercial grade medium components rather than analytical reagents, stainless steel plant rather than glass vessels.

The classical 'scale-up' problems arise from the different ways in which the process conditions change as the scale of the operation increases. A typical example is the relationship between the volume and surface area of a fermentation vessel. The volume is proportional to the cube of the vessel diameter, the wall area is proportional to the square. Thus as the size of the vessel increases, the surface area relative to volume decreases dramatically; this has important implications for heat transfer into cooling jackets for sterilization or temperature control during growth. To handle this problem, it is usual to identify the key parameter in a process, i.e. the one that is the most sensitive to scale-up. In aerobic fermentations this is usually that of maintaining availability of dissolved oxygen. The system is then designed to maintain this 'criterion of scale-up' at the same value at all scales of operation, while other, less important parameters will vary (for details see Winkler, 1983).

Fermentation media

Typically, raw materials account for 70% of the operating costs of an enzyme fermentation. An economic medium is therefore essential and much development work is directed towards the replacement of expensive medium components with cheaper alternatives. The medium should provide all the essential elements required for growth of the microorganism and regulatory molecules (inducers and repressors) should be present or absent as appropriate.

Medium components Those microorganisms used for the production of enzymes require organic compounds as a source of carbon and energy. Carbon is the most abundant element by weight in microbial cells comprising approximately 50% of the biomass and the carbon source (which is usually also the energy source) is therefore the principal ingredient of the culture medium. Refined sugars are too expensive to use on a commercial scale, particularly for the manufacture of relatively cheap bulk products such as amylases or proteases. Consequently raw ingredients such as agricultural wastes are used, often as a combination of several primary materials to help avoid price fluctuations and scarcities due to crop failures. A common carbon source is molasses derived from the sugar industry. Milk whey, obtained as a by-product of the cheese industry, is useful for microorganisms that utilize lactose of which it contains some 68% as a dried powder. Other common carbon sources include various cereals, particularly barley, maize (corn) and rice. Malted barley in which germination has been initiated and then halted by kilning contains partially hydrolysed starch and protein and generally supports improved growth yields, but it is more expensive.

Nitrogen is generally supplied as ammonium salts but growth and product yield will usually be increased by the addition of organic nitrogenous compounds. Corn steep liquor, a by-product of maize starch production, contains 24% protein of which much is in the form of amino acids. Other nitrogenous materials used in commercial media include meals derived from blood, meat, fish, peanuts and soybeans; the choice will depend on market prices and suitability for the process in hand. These carbon and nitrogen sources are fairly rich in purine and pyrimidine bases, vitamins and other growth factors and it is unlikely that it will be necessary to add such compounds.

Of the mineral salts required, phosphate and sulphate are generally added in quite large amounts (about 1 and 0.5 g l^{-1} respectively) and it may be necessary to add smaller amounts of metal ions such as Mg^{2+}. Trace elements will be amply supplied from the other materials. Strong buffers, mostly phosphate, are sometimes used to reduce pH fluctuations during growth if pH control is not applied.

Medium formulation Optimization of growth medium composition for maximum product yield is not an easy task. Traditionally, an empirical approach has been adopted in which the effects of individual medium constituents on growth and product yield are assessed in shake-flask cultures. Carbon sources are examined, then nitrogen sources, until every combination of components has been studied and an optimized medium reached. During this process, repressing compounds will be discarded and those molecules that stimulate or induce enzyme synthesis will be retained. This has been successful but is time-consuming and involves a very large number of analyses. Moreover, since ingredients are examined individually, interactive effects between components are often missed.

An alternative approach is based on process optimization techniques used in chemical engineering (Isaacson, 1970). The primary purpose of this approach is to screen a large number of medium components for their effect on product yield so that the most important can be identified and studied in greater detail. By combining the components in high or low (absent if non-essential for growth) concentrations according to a statistically devised (fractional factorial) design, a large number (n) of components can be examined in n-1 cultures. Thus the effect of an individual component on enzyme yield can be assessed in a relatively small number of experiments, even though several components will vary in each culture. This approach has been successful for the determination of mineral salt and metal ion requirements once the carbon and nitrogen sources have been selected by traditional procedures (Ingle & Boyer, 1976).

A recent approach to medium design, particularly suitable for the choice of carbon and nitrogen sources, involves continuous culture in a chemostat. Briefly, a chemostat culture in steady state is challenged with pulses of individual medium components. If the cell concentration shows a transient increase, then the medium is limiting for that particular substrate and the concentration in the medium should be increased to obtain maximum biomass yield. Inducers and repressors can be identified from transient increases or decreases in product yield upon addition of particular molecules. Thus key components of a medium can be rapidly assessed without recourse to large numbers of individual shake-flask experiments.

Sterilization Most media are sterilized in the fermenter as a batch operation by heating coils or a jacket with steam, or by injecting steam directly into the vessel. It may require several hours to reach temperature (120°C) and cool again with severe loss of production time. Moreover, keeping complex media at a high temperature for long periods leads to medium deterioration (e.g. Maillard reactions between amino acids and sugars, denaturation of proteins) and poor growth. Continuous sterilization is therefore gaining wide acceptance. Continuous heat sterilization involves plate or spiral heat exchanges which heat the medium to very high temperatures (150°C) for a few minutes. Heat economy can be obtained by using the hot sterile medium to warm up the incoming medium, and the short process time preserves the nutritive value of the medium. Such sterilization plants are expensive and generally used to service several fermenters. The fermenters must be sterile to receive the medium and this is generally achieved by direct sparging with steam. Continuous sterilization is, of course, essential for continuous culture systems.

Preparation of inoculum

Since sterilization on an industrial scale is seldom absolutely perfect, it is necessary to inoculate the vessel with a large, vigorous inoculum in order that the desired microorganism can compete with and repress any contaminants. Moreover, the number of propagation stages should be minimized in order to reduce the risk of contamination and preserve the production capability of the strain. The organisms used are generally highly mutated and several subcultures could lead to reduced yields from genetically unstable strains. A typical scheme would therefore involve inoculating a shake flask from a lyophilized stock culture. During mid- to late-exponential phase, this would be used to inoculate a small fermenter (300 to 500 litres) which would probably contain medium that resembled the final production medium. Again, before the stationary phase was reached (10 to 80 h depending on the process), this would be used to inoculate the production vessel which would be of 3000 to 5000 litres capacity. If very large vessels (50,000 litres) were to be used, a second seed fermenter would be necessary.

At each transfer rigorous control is needed to avoid three major causes of failure. Contamination may be detected by frequent microscopic examination or preparing spread plates but the latter really provides an historical record only. Infection by bacteriophage is generally revealed as slow growth with concomitant reduction of yield. The proliferation of low-yielding mutants is difficult to control, but fortunately rare. Thus at each stage, enzyme yield, biomass, pH and the presence of contamination should be monitored and the inoculum discarded if necessary.

Process conditions

Batch culture   Industrial enzymes are almost invariably produced by batch culture methods. Genetic instability of production strains, problems with contamination and difficulties obtaining a stationary phase product from exponential phase cells have severely limited the application of continuous culture. A diagram of a typical enzyme fermentation plant is shown in Figure 17. It comprises a stainless steel vessel of 10,000 to 50,000 litres capacity with associated equipment. After inoculation, the fermentation will proceed for 30 to 150 hours depending on the enzyme and microorganism involved. Enzyme activity is monitored during this time but, because assays are often time-consuming, characteristic shifts in some other parameter such as pH or oxygen demand are generally used to assess the optimum point for harvesting. Moreover, enzyme activity is not the sole criterion to be considered in determining the optimum process time; the properties of the broth with regard to product extraction and efficient utilization of the plant capacity must also be considered.

Facilities are usually available for monitoring and controlling parameters such

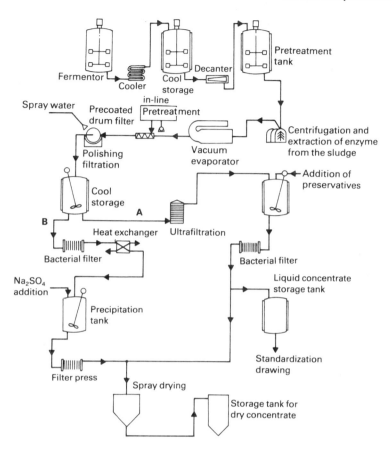

Fig. 17 Schematic diagram of a typical enzyme fermentation process. Examples of useful measurements and controls are indicated (from Aunstrup et al., 1979).

as oxygen tension (aeration and agitation) foaming, pH and temperature. Aeration is achieved by direct injection of air under pressure. The air is first sterilized by filtration, perhaps with a heat pretreatment involving compression and partial decompression. Oxygen solubility is low in aqueous media and decreases with increasing temperature. Oxygen tends, therefore, to be the critical factor in an aerobic 'fermentation', for example with *Bacillus* or *Aspergillus*. Aeration is inseparable from agitation which accelerates oxygen transfer into the aqueous phase by disrupting bubbles thus increasing their relative surface area. It also introduces turbulence which delays the rise of the bubbles to the surface. If the culture has a high viscosity, oxygen transfer is impeded but can be improved by introducing baffles which further retard the rise of the bubbles. The volume of air required is great, often of the order of the volume of the fermenter per minute. Thus the bubbles occupy a considerable proportion of the medium volume, perhaps 20%, a factor that must be considered in the design of the system. With the large volume of air, intense agitation and rich medium, foam becomes a problem. Foam sensors can be fitted to the vessel which control the addition of antifoam which may be vegetable or animal oils or a silicone oil.

Microbial metabolism will affect the pH of the culture and some form of control is therefore essential. The cost of buffers (such as phosphate salts) is very expensive on the commercial scale and automatic pH control is used with hydrochloric acid, sodium hydroxide or ammonia addition. Finally, microbial growth is exothermic and a considerable amount of heat must be removed to maintain the correct temperature. Cooling jackets through which cold water is circulated are often preferred to interior coils because the latter take up valuable space, provide additional surfaces to which microorganisms can adhere and are difficult to clean. However, jackets have a lower heat transfer coefficient than coils and in very large vessels coils cannot be avoided. In general, jackets and coils give similar specific heat transfer rates with vessels of about 3 metres diameter, the size of a typical production-scale stirred tank fermenter. Above this size coils are required.

Extended feed and continuous culture It is sometimes necessary to provide very high substrate concentrations to achieve maximum enzyme yield. For example, in a patented process for the production of glucoamylase, 25% (w/v) liquefied starch is used in the medium. In many cases this leads to problems with catabolite repression, inhibition of growth due to high osmotic pressure and viscosity. These difficulties may be overcome by continuous feeding of carbohydrate or protein. The feeding programme may be governed by monitoring changes in pH or some other suitable parameter or may be established empirically.

In continuous operations, fresh medium flows into the vessel and product flows out continuously. These may be of the homogeneous, fully mixed format or the heterogeneous plug-flow type as described for enzyme reactors in Chapter 1. Continuous systems have found little application in enzyme manufacture, the best-known example being glucose isomerase production from *Bacillus coagulans*. This fermentation is carried out at about 50°C under combined oxygen and glucose limitation and maximum productivity is usually maintained for more than 200 hours. The principal disadvantage of batch operations is the high proportion of 'down' time needed to empty, clean, sterilize and recharge the fermenter. Although continuous operations avoid this and can run for several weeks or even months, the continual inflow of new medium presents a high risk of contamination. This is accentuated since the producer organism is usually highly mutated, grows relatively weakly and is rapidly displaced by an incoming contaminant. Stability of product yield also presents problems. Thus continuous fermentation is best suited to a growth-associated process with a low risk of contamination and a genetically stable organism such as single-cell protein production.

Enzyme extraction and purification

Although industrial enzymes are sometimes sold as crude preparations of the culture broth, most are extracted and purified to a certain degree. This achieves several aims; toxic or undesirable metabolites and microorganisms are removed, standard activity, quality and stability can be maintained, and an acceptable odour and colour can be produced. The main difficulty originates from the variable composition of the fermented broth which often contains large amounts of colloidal material and can be highly viscous. The process generally uses relatively simple procedures such as centrifugation and precipitation, although more complicated techniques are sometimes needed when more stringent purity is required, particularly of intracellular enzymes.

Extracellular enzymes At the end of the fermentation, the broth is rapidly cooled to 5°C to promote stability of the product and restrict growth of any contaminating microorganisms. The pH is adjusted to the optimum for the stability of the enzyme if necessary, and the solid materials, microorganisms and particulate matter from the medium, are removed. Fungal biomass can usually be harvested by direct centrifugation of the culture, but the smaller size of bacteria requires pretreatment of the culture by flocculation to effect an efficient separation in the centrifuge. Bacterial or yeast cells in suspension at neutral pH, or thereabouts, carry a negative charge because of phosphate and carboxyl groups on the wall. Flocculating agents act by neutralizing this charge and forming large clumps of cells that drop out of suspension more readily; aluminium sulphate and calcium chloride come in this category. Bacteria may also be flocculated without charge neutralization by certain synthetic anionic or non-ionic polyelectrolytes often based on acrylates and ethyleneimine. Flocculation is attractive because it is inexpensive, rapid and allows cheaper centrifuges to be used.

After pretreatment of the broth it is centrifuged. Since industrial centrifuges generally have a restricted radius of rotation (mechanical stress increases with the square of the radius of rotation and safety limits are rapidly reached) and the need for continuous flow-through of liquids restricts the angular velocity available, they are less efficient at sedimenting particulate material than laboratory-scale machines. Centrifugation is also hindered by viscous fermentation broths and particles which are often small (although increased by flocculation) and of low density. One approach to centrifuge design to enable rapid sedimentation is to spread the liquid layer over a large surface area in a thin film. Two common machines work on this principle. The multichamber centrifuge has, as the name suggests, multiple chambers in which the liquid layer is thin. The disc centrifuge contains a stack of cone shaped discs in the bowl. Fluid runs over these while the solid material is deposited on the surface and subsequently collects in the perimeter of the bowl. Both batch and continuous discharge of the harvested material are available, the latter allows the machine to be run for long periods without interruptions to clean the bowl.

An alternative to centrifugation is filtration. The effectiveness of a filter is severely reduced as fine particulate material aggregates on the surface and presents a high resistance to the liquid to be filtered. To combat this, filter aids such as diatomaceous earths are used which retain the smaller particles and form a 'cake' on the surface of the filter that can be readily removed. A common form of industrial filter is the plate and frame press which comprises filter cloths trapped between corrugated plates. Fluid passes in at one side of the cloth and out via the corrugations to a collection pipe. A disadvantage is that the plates have to be disassembled to clean the filter cloths periodically. In the rotary drum filter, vacuum is applied to the inside of a fabric-covered, hollow drum rotating horizontally in a trough that contains the fluid to be filtered. The drum is precoated in a filter aid and as it rotates, the cake is scraped from the surface of the filter by a metal blade. Thus a clean filter surface is continually presented to the fluid and clogging of the filter is prevented.

The clear enzyme solution is concentrated by vacuum evaporation or ultrafiltration. During evaporation the temperature should be maintained below 35°C to prevent denaturation of the enzyme. Ultrafiltration and reverse osmosis are relatively recent innovations for the enzyme industry. When high pressure is applied to a solution in order to force solvent (maximum molecular weight about 250 daltons) across a membrane from a strong solution into a weak one, the process is referred to as reverse osmosis. With a larger pore size membrane (500 to 300,000 dalton maximum depending on the process) the hydraulic force results in the passage of solute molecules too, and the system is called ultrafiltration. There are two types of ultrafiltration membrane. The microporous membrane is rigid with small pores running through it. Very small molecules will pass through; larger ones will be retained at the surface. Intermediate size molecules will be lodged in the membrane and ultimately block the pores. These filters have been superseded by the diffusive membrane, which is essentially a homogeneous hydrogel membrane through which solvents and solutes are transported by molecular diffusion under the action of a concentration gradient. Since the membrane contains no pores in the conventional sense, it does not block in the same way as microporous membranes (Melling & Phillips, 1975).

The concentrated enzyme may be clarified by a 'polishing' filter and the remaining microorganisms removed by filtration through cellulose or asbestos filter pads. For liquid enzyme preparation, the concentrate is diluted to the appropriate concentration with water, stabilizers are added such as sodium chloride, benzoates or sorbates, and packaged. If a dry product is required, the enzyme may be precipitated with organic solvents (acetone or ethanol) or inorganic salts (sodium sulphate or ammonium sulphate) to obtain a higher degree of purity. The precipitate is dried, ground and mixed with an extender (starch, dextrins or sodium chloride) to the desired activity. Alternatively, the concentrated enzyme may be spray-dried and packaged after standardization of activity.

It may seem to the biochemist that these preparative procedures are crude, but for the majority of commercial enzymes purity is a minor consideration. Most microorganisms are unlikely to secrete more than twenty proteins and the fermentation is designed to maximize synthesis of one enzyme wth concomitant repression of others. These simple procedures will, therefore, provide a relatively pure product and, for many uses a degree of 'contamination' by protease or amylase is very useful. However, if pure enzymes are required the concentrated enzyme can be further purified, generally by a chromatographic procedure (Bucke, 1983).

Intracellular enzymes The biomass is removed from the cooled fermentation broth by centrifugation, since this avoids the subsequent removal of filter aid, and washed to removed medium components. Breakage of the cells is achieved by a variety of processes depending on the microorganisms. Ball-milling, which involves shaking cells with small glass beads in a rapidly vibrating vessel, is an excellent means of breaking most microbial cells. Sonication, freezing and thawing, liquid shear obtained by forcing a cell suspension through a small orifice at high pressure, and enzymic removal of the cell wall all have their uses for various systems (Bucke, 1983). Cell debris is removed by centrifugation, and nucleic acids are either hydrolysed with enzymes or precipitated with high molecular weight cations such as polyethyleneimine or streptomycin sulphate. Since many other proteins will be present, purification by chromatography is often necessary. Because of the complexity and expense of these operations, every effort is made to avoid purification and to use intracellular enzymes immobilized within the cell.

Finishing of enzymes Enzymes, like most other proteins, are antigenic, and considerable problems occurred in the late 1960s through workers breathing enzyme dusts which provoked allergic responses. To eliminate the dust from powdered enzyme products, the particle size was increased from around 10 μm to between 200 and 500 μm diameter by 'prilling' or 'marumizing'. The former involves mixing the enzyme with polyethylene glycol or something similar, and preparing small spheres by atomization. A marumized granulate is formed by mixing the enzyme with a binder and water, and extruding it as long filaments. These are converted into spheres by the marumizer, dried and covered with a waxy coat. All detergent enzymes are granulated, but the additives used to form the granules are often inappropriate for use in foods. Liquid enzyme concentrates avoid these problems.

Regulation and legislation

Microbial enzymes are natural products that have traditionally been used in foods, and for many years authorities felt that legislation was unnecessary. However, the allergic responses involved with the introduction of enzyme-containing washing powders brought enzymes into the mass media, and the public generally viewed them as a potential health hazard. At the same time, new enzymes were marketed that were produced from microorganisms other than those traditionally involved with food, and many governments felt that there was a need for some form of legislative control.

Safety It is generally agreed that enzymes derived from animals or plants that would normally constitute food, or from microorganisms associated with food, can be regarded as safe, and no safety evaluation is necessary. For enzymes from microorganisms outside this category, safety testing is needed. Three areas of potential concern have been identified; the catalytic activity of the enzyme itself, allergic reaction caused by any proteins present in the product and the presence of toxic metabolites such as mycotoxins or antibiotics. Of these, the first is unlikely to constitute a hazard to health since foods often contain a variety of enzymes and there is no reason to suspect that these might be dangerous. Furthermore, the activity of enzymes used in foods is generally low, and the protein is usually denatured in the final food. Although allergic responses have been mentioned in relation to inhalation of enzyme dusts, there is no evidence to suggest that ingestion of food containing small amounts of added enzymes has given rise to any case of allergy amongst consumers. There is no reason to suppose that enzymes may be any more allergenic than other proteins, so this too is considered unlikely to constitute a hazard to the health of the consumer. The most difficult hazard to exclude is that of potentially toxic metabolites. These may be present as contaminants of the raw materials, they may be produced by contaminating microorganisms, or by the enzyme-producing organism. Good manufacturing practice should exclude the first two of these; the third is the main concern and can only be evaluated by toxicity tests. The minimum test requirement will usually be a 90-day oral feeding study in a rodent species and a non-specific biological screening test for toxins but different countries have their own regulations which are often more stringent than this (Denner, 1983).

Specifications of purity Each batch of enzyme will have to meet a set of safety specifications for the grade of product. Broadly speaking there are four grades: analytical, pharmaceutical, food and technical. The safety specification for a food grade enzyme gives limits for arsenic, lead, heavy metals, aflatoxin and antibiotics and for certain microorganisms namely *Salmonella, Escherichia coli,* coliforms and *Pseudomonas aeruginosa* (Denner, 1983).

Summary

Extracellular enzymes are generally prepared from microorganisms grown in submerged culture. These industrial scale processes are referred to as fermentations even though the culture conditions are often highly aerobic. Scale-up involves translating the laboratory process into an industrial size operation, and problems may arise from the large quantities involved. Alternatively, problems originate within the process itself such as the difficulty of maintaining efficient aeration in a very large fermenter. This results in the industrial scale process often being quite different from the original laboratory scale fermentation. Commercial media comprise inexpensive raw ingredients, often agricultural surpluses and wastes such as corn steep liquor, cereals and milk whey. Media and process conditions are optimized to stimulate maximum product yield concomitant with economic utilization of raw ingredients and ease of product recovery. Most enzyme fermentations are batch culture operations because yields tend to be low in continuous culture and the process is susceptible to contamination. Sometimes it is advantageous to feed the carbon source into the fermentaton over an extended period thus avoiding catabolite repression.

Recovery of extracellular enzymes is straightforward. Biomass is removed by centrifugation; bacteria generally require pretreatment by flocculation. The supernatant fluid is concentrated, filtered a second time, diluted to a standard activity, stabilized by the addition of sodium chloride or sorbate and packaged. Enzyme powders are precipitated or spray dried and, to avoid dusts converted into small waxy-coated spheres. All enzyme products are subject to regulations, but those that are derived from organisms associated with natural foods are deemed safe. However, enzymes from microorganisms not usually associated wth foods require extensive toxicity tests before they can be used for food manufacture. All enzyme products must meet safety specifications before they can be marketed.

References

AUNSTRUP, K., ANDERSEN, O., FALCH, E. A. and NIELSEN, T. K. (1979). Production of microbial enzymes. In: *Microbial Technology,* second editon, Vol. 1, pp. 281–309. Edited by D. Perliman, Academic Press, New York and London.

BUCKE, C. (1983). The biotechnology of enzyme isolation and purification. In: *Principles of Biotechnology,* pp. 151–71. Edited by A. Wiseman. Blackie and Son Ltd., Glasgow.

DENNER, W. H. B. (1983). The legislative aspects of the use of industrial enzymes in the manufacture of food and food ingredients. In: *Industrial Enzymology: the Application of*

Enzymes in Industry, pp. 111–37. Edited by T. Godfrey and J. Reichelt. The Nature Press, New York.

INGLE, M. B. and BOYER, E. W. (1976). Production of industrial enzymes by *Bacillus* species. In: *Microbiology 1976*, pp. 420–26. Edited by D. Schlessinger. American Society for Microbiology, Washington, D.C.

ISAACSON, W. B. (1970). Statistical analyses for mulivariable systems. *Chemical Engineering*, June, 69–75.

JANSEN, J. C. and HEDMAN, P. (1982). Large scale chromatography of proteins. *Advances in Biochemical Engineering* 25; 43–99.

MELLING, J. and PHILLIPS, B. W. (1975). Large-scale extraction and purification of enzymes. In: *Handbook of Enzyme Biotechnology*, pp. 58–88. Edited by A. Wiseman. Ellis Horwood Ltd., Chichester, U.K.

WINKLER, M. A. (1983). Application of the principles of fermentation engineering to biotechnology. In: *Principles of Biotechnology*, pp. 94–143. Edited by A. Wiseman. Blackie and Son Ltd., Glasgow.

Overall summary

This book is a blend of pure and applied microbiology and deals with the bio-technology of industrial enzymes. Most of these enzymes are extracellular proteins and Chapter 2 covers recent biochemical and genetic studies aimed at elucidating the process of protein transport across membranes. This combined approach has been outstandingly successful. The signal hypothesis proposed that secreted proteins are synthesized as precursors bearing an NH_2-terminal peptide that directed the protein into and across the membrane. This has been validated by characterization and sequence analysis of various signal peptides, which have led to proposals as to how the peptide might function. Genetic studies have revealed a secretory machinery in the plasma membrane of *Escherichia coli* which perhaps 'drives' the secretion process. Understanding of protein secretion has developed much quicker than analysis of the regulation of extracellular enzyme synthesis covered in Chapter 3. Induction and repression of enzyme synthesis and the regulation of metabolism by carbon and nitrogen availability are known to occur but we understand little of the molecular mechanisms involved in organisms other than *E. coli*.

The types of enzymes produced from microorganisms and their industrial applications are covered in detail in Chapter 4. Of particular commercial import-ance are the serine proteases of *Bacillus* species which are included in household washing detergents and the acid proteases of some filamentous fungi which are replacing calf rennet in cheese manufacture. The amylases of *Bacillus* and *Aspergillus* species are used to hydrolyse starch into glucose which is subsequently converted into fructose using glucose isomerase. Since fructose tastes very sweet, these syrups are being used for sweetening foods and beverages in place of sucrose. An important development, that allowed the commercial exploitation of the relatively expensive glucose isomerase, was enzyme immobilization on solid surfaces and the introduction of enzyme reactors.

With this background, methods for screening strains for the secretion of par-ticular enzymes and strain development are covered in Chapter 5. Of particular importance is the application of genetic engineering techniques which is enabling not only the construction of strains that produce exceptionally high yields of particular proteins but promises the prospect of engineering almost any protein such that it will be secreted by the host organism. The final chapter returns to industrial practice and considers the problems involved in manufacturing a bulk microbial product on a commercial scale. Medium design, plant operation and the processing of the product are covered, as well as the safety and legislative aspects of enzyme manufacture.

Glossary

Attenuation: the termination of transcription due to the formation of a stem-and-loop structure (terminator) in the leader sequence of mRNA.

Chitin: a polysaccharide, closely related structurally to cellulose and composed of a linear array of β-linked 2-acetamido-2-deoxy-D-glucose units.

Cotranslational secretion: the process by which a nascent polypeptide is transferred across a membrane as it is translated.

Docking protein: a protein in the membrane of the endoplasmic reticulum that cancels the effect of the signal recognition particle and allows cotranslational secretion to proceed.

Endo-acting: describes enzymes that attack bonds at random in a macromolecule.

Exo-acting: describes enzymes that hydrolyse bonds progressively from the end(s) of a macromolecule.

Fermentation: those energy-yielding metabolic pathways in which organic compounds act as both electron donor and electron acceptor. Used in industrial circles to describe any large-scale growth of microorganisms either aerobic or anaerobic.

α-Helix: secondary structure of a protein in which hydrogen bonds formed between the components of the peptide linkages are nearly parallel to the long axis of the helix. One turn occurs every 3.6 amino acid residues.

Leader sequence: a region of about 200 bp preceeding the first major structural gene of an operon controlled by attenuation. Also used to describe the coding region for the NH_2-terminal peptide extension of secreted proteins (see signal peptide).

Limit dextrins: products remaining after exhaustive hydrolysis of amylopectin or amylose by an amylase. Term may also be used for other polysaccharides and their enzymes.

Linkages: inter-sugar linkages may be of the α or β form depending on the anomeric configuration of those carbon atoms through which the linkage is effected. Cellobiose a 1,4-β-, and maltose, a 1,4-α-disaccharide of glucose are examples.

Cellobiose   Maltose

Liposome: lipid vesicles comprising an aqueous compartment bounded by a lipid bilayer.

Liquefaction: reduction of starch to oligosaccharides of about 10 residues in length by enzymic or acid hydrolysis.

Operator: a DNA sequence that stimulates or represses transcription of adjacent genes by binding an activator or repressor protein respectively.

Extracellular Enzymes

Outer membrane: the surface membrane of Gram negative cells.

Peptidoglycan: a net-like polymer comprising polysaccharide 'backbones' linked by short peptides. It provides structural rigidity to the bacterial cell wall.

Periplasm: the hydrophilic zone bounded by the cytoplasmic (inner) and outer membranes of a Gram negative bacterium.

Polysome: several ribosomes concurrently translating a mRNA.

Post-translational secretion: transport of a completed protein across a membrane.

Preprotein: the precursor form of a secretory protein in a eukaryotic cell (see proprotein).

Promoter: a sequence of DNA at which RNA polymerase binds preferentially and within which RNA synthesis is initiated.

Proprotein: term used for the precursor form of secretory proteins in *Escherichia coli*. Also refers to the precursor form of a functional protein in eukaryotes which is not necessarily a secreted protein. In the case of secreted proproteins, e.g. insulin, the secretory precursor is preproinsulin.

Protoplast: an osmotically fragile cell from which the wall has been completely removed. Generally refers to Gram positive bacteria treated with lysozyme.

Saccharification: enzymic or acid hydrolysis of starch (or more usually liquefied starch) to maltose, maltotriose and other low molecular weight maltooligosaccharides.

Sigma (σ) factor: a polypeptide component of RNA polymerase responsible for initiation of transcription at specific sites on the genome.

Signalase: see signal peptidase.

Signal peptidase: a protease that removes the signal peptide from a precursor protein. Also called leader peptidase.

Signal peptide: an extension of 15 to 30 amino acids at the NH_2-terminal end of a protein that interacts with the cytoplasmic membrane and directs the protein into and/or across the membrane.

Signal recognition particle: a complex of six polypeptides and an RNA molecule that blocks translation once the signal peptide has emerged from the ribosome (see docking protein).

Signal sequence: those codons in the mRNA responsible for synthesis of the signal peptide. Also used for signal peptide.

Specialized transducing phage: a bacteriophage that has excized incorrectly from a bacterial chromosome and contains a segment of bacterial DNA incorporated in its genome.

Sphaeroplast: an osmotically fragile structure, usually derived from a Gram negative bacterium from which the wall has been partially removed.

Suppressor mutation: intergenic suppression describes a mutation in a second gene that restores a normal phenotype to a mutant organism. Intragenic suppression refers to a second mutation in the gene harbouring the primary mutation.

Transposon: a DNA element which can insert into the bacterial or host chromosome independently of the host cell recombination system.

Transformation: the uptake, and usually inheritance and expression, of exogenous DNA.

Index